30-SECOND
AI AND ROBOTICS

30-SECOND
AI AND ROBOTICS

50 key notions, characters, fields and
events in the rise of intelligent machines,
each explained in half a minute

Editor
Luis de Miranda

Contributors
Sofia Ceppi
Neha Khetrapal
Ayse Kucukylimaz
Pedro U. Lima
Andreas Matthias
Lisa McNulty
Luis de Miranda
David Rickmann
Mario Verdicchio

Illustrations
Steve Rawlings

IVY PRESS

First published in the UK in 2019 by
Ivy Press
An imprint of The Quarto Group
The Old Brewery, 6 Blundell Street
London N7 9BH, United Kingdom
T (0)20 7700 6700 **F** (0)20 7700 8066
www.QuartoKnows.com

British Library Cataloguing-in-
Publication Data
A catalogue record for this
book is available from the
British Library.

ISBN: 978-1-78240-547-4

This book was conceived,
designed and produced by
Ivy Press
58 West Street, Brighton BN1 2RA, UK

Publisher **Susan Kelly**
Creative Director **Michael Whitehead**
Editorial Director **Tom Kitch**
Project Editor **Caroline Earle**
Designer **Ginny Zeal**
Picture researcher **Katie Greenwood**
Glossaries **Luis de Miranda**

Printed in China

10 9 8 7 6 5 4 3 2 1

CONTENTS

INTRODUCTION
Luis de Miranda

When personal computers started appearing in homes more than thirty years ago, many non-users thought they were complicated devices for a minority of engineering-minded specialists – objects that the average citizen could safely ignore. Not so. In fact, the happy few who tried to understand what was going on in the digital world soon developed a clear advantage: they became adapted to the new world, and others then had to catch up. Today, artificial intelligence and robotics are spreading like technological wildfire. If you ignore the phenomena, you are simply burying your head in the sand. Yet fear of these technologies is understandable, given the information overload about the latest achievements of AI and robotics.

The title you are holding was carefully conceived by a team of highly qualified specialists as the gentlest way to explain to non-specialists what this new technological revolution is about. It is a comprehensive guided tour of what it means for us humans to be surrounded by machines that can outsmart us, replace us, or mimic us more and more convincingly. This book is your step towards embracing or even changing the new technologies, or at least having an informed and critical opinion about them.

30-Second AI and Robotics is not about the future – it's about the present. AI is already capable of booking a table for you by calling the restaurant and having an effective short dialogue without being identified as a machine. Neural connections can be made between a human brain and a robotic device so that people who have lost the use of an arm can utilize the power of mental focus to grab a drink with their mechatronic limb. We are already sending robots into the cosmos to prepare for the space colonies of tomorrow. Machines are helping elderly people or disabled children to improve their quality of life.

In some storage facilities, robotic swarms have almost completely replaced human workers, and a driverless car or truck might have crossed your road without you noticing it.

This book is an informative cultural and intellectual story. It is divided into seven chapters, which take off from historical myths regarding robots and AI, fly through the main ideas regarding intelligent machines in society, and land on what seems to be a likely scenario for the future: our anthrobotic merging with machines, with which we will live symbiotically. On this journey, you will discover the Turing Test, deep learning, drones and medical AI. You'll read about smart assistants, virtual reality, the Internet of Things, augmented minds, neural networks and all areas of the so-called fourth industrial revolution.

Each entry discusses one of the 50 major ideas, inventions, tools or applications of AI and robotics. Every topic is presented in the same intuitive format: the 3-Second Byte is a quick summary; the 30-Second Data section offers a comprehensive overview of the subject with an explanation of how it works; and the 3-Minute Deep Learning paragraph presents a topic with ethical or philosophical implications for you to reflect on. The links to related subjects allow you to read the book in a non-linear way, following related themes. The 3-second biographies are inspiring micro-portraits of leading figures such as Elon Musk, Isaac Asimov, Alan Turing and Geoffrey Hinton. Each chapter contains a glossary of the most important terms.

Whether you are a student wondering about your future career in this anthrobotic world or are simply curious about our new social reality, this book is the easiest way to become informed without having to learn how to code or to connect a single electric wire. AI and robotics are a social-cultural phenomenon, sometimes positive, sometimes potentially negative, a series of ideas and practices that are shaping our world. Should we be afraid of robots and AI? This book will help you to make up your mind.

FICTIONAL & HISTORICAL ROBOTS

FICTIONAL & HISTORICAL ROBOTS
GLOSSARY

AI winter Artificial Intelligence is a field that experiences cyclical periods of general interest and funding. In between the periods of enthusiasm for AI are 'wintry' years during which research appears to stagnate. During such periods, it can be beneficial to pause, reflect and generate new ideas.

analogue A term used to describe non-digital technology, in which signals are not converted into discrete digits of 0 and 1, but rather into continuous physical pulses. A mechanical watch is analogue, and so is the human voice in air.

android Coined from the Ancient Greek word meaning *male* and *oid* (meaning *like* or *having the form of*), 'android' is a word for robots that are male or human-like. A robot with a female appearance should be referred to as a *gynoid*, although the term android is often used instead.

artificial neural networks AI computing systems that are designed to work like a brain. They learn progressively by connecting inputs of information with outputs, through a filtering process that is partly autonomous, and via a collection of nodes called artificial neurons, in which each connection is a simplified version of a synapse. This model is responsible for the success of deep learning since 2011.

autonomous In the field of robotics and AI, 'autonomous' means being able to function with limited human intervention. An autonomous system can adapt its behaviour depending on the situation.

cybernetics Norbert Wiener defined cybernetics in 1948 as the study of control and communication in systems involving machines and humans. This interdisciplinary field is concerned with organization, efficiency and governance. It tends to see all living beings as part of systems, and systems themselves.

cyborg A cybernetic organism, a hybrid being composed of flesh and mechatronic parts. A cyborg could be a bionic robot that has human features, or a human enhanced with electronic and mechanical parts.

humanoid A less gendered term for android. 'Humanoid' can also describe a robot that has some kind of human feature, without being as human-like as an android or a gynoid. Most engineers believe that a humanoid robot is more easily accepted by humans. The same reasoning does not seem to apply to AI research: most AI systems today are expected to process information better than humans, and in ways that do not have to mimic our reasoning.

mechatronic machines Machines made of mechanical and electronic parts, and computer elements. Mechatronic engineers work on solving the integration of these various parts. Today, a fourth component is added to the mix: biological elements. A new interdisciplinary study has emerged that is probably the future of robotics: biomechatronics.

positronic Author Isaac Asimov imagined the positronic brain to describe a form of robotic minimal consciousness built upon his Three Laws of Robotics (see page 144). The positron is an antiparticle that is the antimatter counterpart of the electron.

psychohistory A fictional science imagined by Isaac Asimov in his novel *Foundation*, which would combine history, sociology and statistics in order to predict the future of civilizations. Even though individuals are unpredictable, applying the laws of statistics to large groups could allow scientists to predict future events. Today, some researchers take this idea seriously in the field of social physics.

uncanny valley This idea, proposed by robotics professor Masahiro Mori, suggests that when humanoids become very human-like, yet not fully so, the general human response is a strong rejection because of their awkwardness. This hypothesis has not been proven, and it might be that humanity becomes used to the spooky appearance of humanoids.

THE GOLEM
& MAGIC

the 30-second data

3-SECOND BYTE
The Golem is the myth relating to an artificial creature brought to life with magic; once granted its strength the Golem was no longer subservient and had to be destroyed.

3-MINUTE DEEP LEARNING
From Caribbean zombies to *2001*'s super-computer HAL, the absence of human-like emotions, breakdown of language or the ultimate lack of control over the artefact are all common features used in 'artificial man' stories. Mary Shelley's *Frankenstein* – the book itself being a modern version of the Golem – inspired the term 'Frankenstein complex', coined by Isaac Asimov to describe the fear of mechanical men.

Before there was a science of AI, people believed it was up to God and the devil to help magicians animate their artificial creatures. In the late sixteenth century, the chief rabbi of Prague, Judah Loew ben Bezalel, is said to have made a creature from mud to protect the Jews in his community – the Golem. This story echoes the Biblical creation of man by God, with the word 'Golem' being used to describe Adam, the unfinished human being. According to Jewish folklore, in order to animate and bring life to the raw shape of clay, the creator had to write a magic sign on the forehead of the Golem. This *shem*, one of the Hebrew names for God, provided the life force of the creature, which, after turning against his maker, became impossible to control and ultimately had to be destroyed. The Golem might be perceived as similar to a factory robot in the modern sense; a creature that is strong but mindless. In the end the Golem becomes a paradigm for the hazards of creating something that goes far beyond our intent, and the potential danger of creation without control.

RELATED TOPICS
See also
KAREL ČAPEC & THE FIRST 'ROBOTS'
page 18

ISAAC ASIMOV
page 22

HAL 9000 IN *2001: A SPACE ODYSSEY*
page 24

3-SECOND BIOGRAPHIES
ALBERTUS MAGNUS
C. 1200–80
German alchemist and magician, who was said to have had a very talkative bronze head mounted in his study

MARY SHELLEY
1797–1851
British author of *Frankenstein*, a cautionary tale that can be read as a metaphor about the dangers of uncontrolled science

30-SECOND TEXT
Andreas Matthias

Where's the good in creative powers if we cannot control what we create?

JACQUET DROZ & EIGHTEENTH-CENTURY AUTOMATA

the 30-second data

3-SECOND BYTE
The first 'reprogrammable robot' was a scribe made of thousands of cogs invented by Swiss watchmaker Jacquet Droz, at a time where automata – human-like moving machines – were a fad in Europe.

3-MINUTE DEEP LEARNING
In 1740, Jacques de Vaucanson, whom Voltaire called a modern Prometheus, invented a famous mechanical duck that could rise, flap his wings, and appear to eat, digest and defecate. The age of rationalism was about separating the entire world into parts, and automata presented a technocratic worldview according to which the universe was a machine. This inspired nineteenth-century conceivers of calculating engines, such as Charles Babbage.

The Writer, the Lady Musician and the Draftsman: these sophisticated androids were engineered between 1767 and 1774 by a celebrated Swiss watchmaker, Pierre Jacquet-Droz (1721–90), aided by his son Henri-Louis and his adopted son Jean-Fréderic Leschot. Watchmaking was the leading technology of this age. Since the universe was increasingly seen as a giant clock, conceiving autonomous dolls was seen as imitating God. A complex miniaturized mechanism was hidden inside the body of the automata. The Writer, still functioning today, is made of 6,000 different pieces. He holds a goose feather, which dips into an inkwell. While his eyes seem to intentionally follow his work, he draws letters to formulate up to forty characters, a short sentence that can be preset. The Draftsman used a pencil to draw a portrait of the French King Louis XVI with Marie-Antoinette, but could also be 'reprogrammed' to sketch the figure of other royals. The Musician was an elegant woman playing the organ with her moving fingers. Such automata fascinated early-modern European elites, inspiring philosophers including Descartes or Voltaire, rulers like Emperor Frederik II, or military commanders. Long before robots, they challenged our conception of what it is to be human and skilled.

RELATED TOPICS
See also
METROPOLIS &
THE GYNOID MARIA
page 20

GREY WALTER'S TURTLES
page 26

CAN MACHINES THINK?
page 50

3-SECOND BIOGRAPHIES
JACQUES DE VAUCANSON
1709–82
A famous French inventor of machine tools and lifelike automatons

CHARLES BABBAGE
1791–1871
English polymath. By looking at chess-playing automata, Babbage came up with his 'difference engine', the world's first computer

30-SECOND TEXT
Luis de Miranda

The Enlightenment saw humans as sophisticated machines. Today we tend to believe robots will be more than human.

KAREL ČAPEC & THE FIRST 'ROBOTS'

the 30-second data

3-SECOND BYTE
'Robot' is a word that comes from the Czech 'forced work'; it was coined by Karel Čapec in a play about humanoid beings who rebel against their human masters.

3-MINUTE DEEP LEARNING
Human deaths caused by robots or AI-driven machines are relatively rare. The first was registered in 1979 and was of course not intentional. Yet, perhaps to avoid a future rebellion and revenge of humanoid machines, some experts are proposing moral obligations of society towards its machines in the form of robot rights or legal electronic personhood. For example, it is argued that robots should not be sex slaves.

The word 'robot' in its modern sense first appeared in *R.U.R.*, a visionary play written by Karel Čapec (1890–1938) that premiered at the Prague National Theatre on 25 January 1921. R.U.R stood for Rossum's Universal Robots, a label that described artificial humans invented by a mad scientist named Old Rossum. *Robota* meant 'forced labour' and 'slavery' in Czech and Slavic. The word robot quickly became popular around the world as the play was translated and played in dozens of countries by 1923. Each Universal Robot can do the work of more than two humans, thus allowing the latter to stop working and focus on more fulfilling tasks. Rather than mechatronic machines (made of cogs and circuits but no biological component), the robots are sentient androids made of synthetic flesh. Imagined soon after the Russian revolution, *R.U.R*'s humanoids were perhaps seen as a metaphor for the proletarian class: they realize they have feelings and dignity, and that they can be superior to their masters if they unite and revolt. The robots eventually destroy the human species and attempt to replace it as two of them, like Adam and Eve, fall in love with each other. Today, the idea that robots might be, in Darwinian fashion, the new species that comes after us seems less and less fictional to some observers.

RELATED TOPICS
See also
HUMAN-ROBOT INTERACTION
page 78

HUMANOID ROBOTS
page 90

CAN MACHINES HAVE COMMON SENSE?
page 96

3-SECOND BIOGRAPHIES
JULIEN OFFROY DE LA METTRIE
1709–51
French physician and philosopher who wrote the classic *Man a Machine*, in which he claims that humans are sophisticated automata

AUGUSTE VILLIERS DE L'ISLE-ADAM
1838–89
French author of *The Future Eve* and the first to use the word 'android' to describe his fictional robotic woman

30-SECOND TEXT
Luis de Miranda

If you treat robots as slaves, they might thrive to become masters.

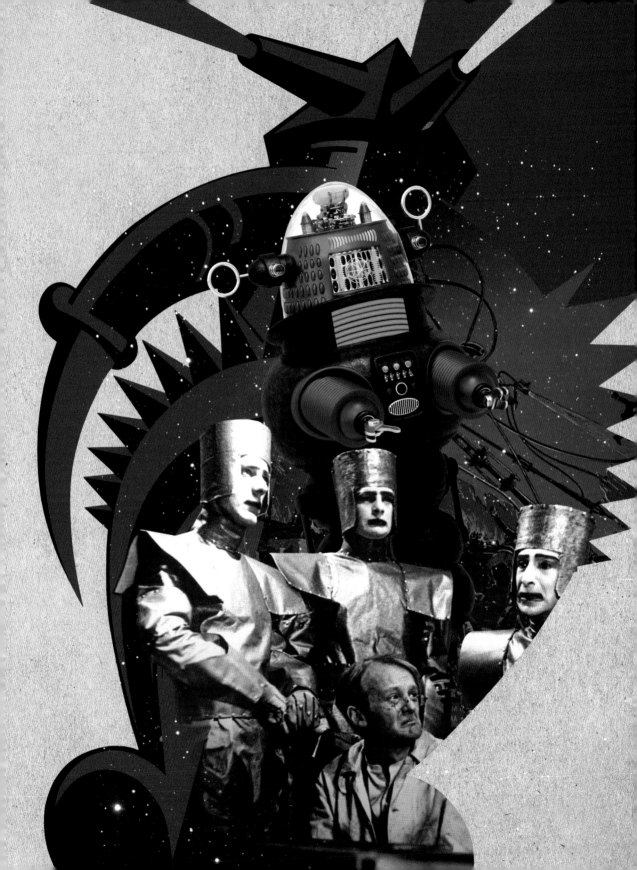

METROPOLIS &
THE GYNOID MARIA

the 30-second data

3-SECOND BYTE
In 1927, Fritz Lang's movie *Metropolis* featured Maria, a.k.a. Maschinenmensch, a metallic automaton with a feminine body – the first famous robot in the history of cinema.

3-MINUTE DEEP LEARNING
In the 1920s, the Western world was experiencing a technological boom in transportation, industry and labour standardization. The theme of the loss of control over our inventions was not new, but the fact that *Maschinenmensch* was feminine, as well as the political activist Maria, sheds some light on the ambiguities of technological and political progress. In some ways, *Metropolis* was also a movie about the fear of women's emancipation.

Fritz Lang, director of the acclaimed *Dr Mabuse* in 1922, was already considered a master of German film when he started shooting *Metropolis* in Berlin. The movie depicts a technological megalopolis where a rich minority lives above the numerous subterranean workers – enslaved and machine-like humans. One day, Freder, the privileged son of the city master Fredersen, falls in love with the activist Maria, a poor young woman who wishes to liberate her fellow workers. Freder becomes aware of the social injustice she is trying to fight and resents his father. Worried, Fredersen asks the mad inventor Rotwang to create a dark double of Maria, in order to fool his son and deceive the workers. The evil plan of the city's master fails, and the workers turn against his totalitarian regime: eventually, they destroy the robot on a pyre. Freder and the real Maria liberate the workers, and injustice is abolished. In the movie, the robotic Maria is half human, half machine, a sort of female Frankenstein's monster. Designed by Walter Schulze-Mittendorff, the robot was played by actress Brigitte Helm. It is never explicitly determined whether the fake Maria possesses feelings, but she can certainly dance and fascinate humans like a femme fatale.

RELATED TOPICS
See also
KAREL ČAPEC & THE FIRST 'ROBOTS'
page 18

GENETIC ENGINEERING & BIOROBOTICS
page 108

3-SECOND BIOGRAPHIES
FRITZ LANG
1890–1976
Austrian-German film-maker, who identified with Expressionism. He deemed the ending of *Metropolis* 'too optimistic'

DONNA HARAWAY
1944–
American professor of science and technology studies, author of the famous *Cyborg Manifesto*, in which she proposes an alliance between feminism and technology

30-SECOND TEXT
Luis de Miranda

Maria, under her metallic shell, appears to have a soul. Is this how we imagined the future of feminism?

January 1920
Born Isaak Ozimov in
Petrovichi, Russia; the
exact date of his birth
is not known

1923
Family emigrates to
United States, and
Isaac learns English
and Yiddish as first
languages, not Russian

1935
Graduates from Boys
High School, Brooklyn,
New York

1942
Publishes 'Runaround' in
the magazine *Astounding
Science*, March 1942

1948
Earns a PhD in
Biochemistry from
Columbia University
School of General Studies

1985–92
Serves as honorary
president of the American
Humanist Association

6 April 1992
Dies in Brooklyn,
New York

ISAAC ASIMOV

Alongside Robert A. Heinlein

and Arthur C. Clarke, Isaac Asimov is considered one of the 'Big Three' writers of twentieth-century science fiction. He grew up in New York, spending much time in his parents' sweet shops. The stores also sold fiction magazines, which Asimov could read for free. Later, he said that these magazines were responsible for his lifelong love of writing. In 1939, Asimov connected with the editor of the magazine *Astounding Science Fiction*, John W. Campbell, who eventually became a friend and a strong influence on the young writer.

Asimov was an excellent student, finishing high school at 15. He became a university biochemistry professor, yet he still found the time to write over 500 books on almost every area of human knowledge, including history, religion, popular science and humour. His books are found in every category of the Dewey Decimal Classification (used to sort books into topics in libraries), except for category 100 – Philosophy and Psychology. This is ironic, since today his Three Laws of Robotics are considered to be one of the earliest and most influential frameworks for robot ethics. He is credited for introducing the words 'positronic', 'psychohistory' and 'robotics' into the English language.

Many of Asimov's robot stories deal with the problems that arise when robots live and work among humans. In a 1942 short story, 'Runaround', he first presented the Three Laws, which have since been both extensively discussed and criticized (see page 144). But his fiction is not limited to robots: he also created the 'Foundation' trilogy of novels about a huge interstellar empire in the future, which became his most famous science-fiction work. He was not only remarkable as an author, but was also a passionate university teacher and letter writer; in the course of his life, he wrote more than 90,000 letters to his fans.

Asimov was a claustrophile: he enjoyed small, enclosed spaces. But his imagination travelled widely. He always resisted specialization, trying instead to escape the boundaries of individual disciplines. Comparing human knowledge to an orchard, he once wrote: 'I looked with horror, backward and forward across the years, at a horizon that was narrowing down and narrowing down to so petty a portion of the orchard. What I wanted was all the orchard, or as much of it as I could cover in a lifetime of running . . .'

Andreas Matthias

HAL 9000 IN *2001: A SPACE ODYSSEY*

the 30-second data

3-SECOND BYTE
This 1960s fictional computer provided a guiding vision to AI researchers for the following fifty years, yet even today we cannot have a credible chat with a machine.

3-MINUTE DEEP LEARNING
The absence of human emotions and the cold rationality of machines are common features of early science-fiction robots. In the 1960s, AI programming was dominated by the idea that thought is logic, and computers were programmed to solve mathematical equations, prove theorems or play chess. After this paradigm failed in the 'AI winter' of the 1980s, research switched to much more human-like cognition using neural networks rather than logic calculus.

The science-fiction author Arthur C. Clarke imagined the all-too-human computer HAL 9000 in his novel *2001: A Space Odyssey* (1968) and he worked with Stanley Kubrick (1928–99) on the screenplay of his film of the same name. HAL, a spaceship computer on an interplanetary mission, can speak and understand speech, recognize faces and steer the ship. He can reason and play chess, understand human emotions and behaviour, and can even lip-read. Like in the Golem story, in Mary Shelley's *Frankenstein* and countless other tales about artificial beings, in the end HAL turns against the human crew and tries to kill them. In a dramatic climax to the story, the last surviving astronaut manages to dismantle HAL, causing the computer to lose his abilities of thought and speech one by one. HAL (in the film) was supposed to be built in 1992 (1997 in the book). Today's computers have some of HAL's abilities, but are still far away from being able to converse freely as he did. HAL expresses the optimism of the 1960s AI community: one of the advisers for the movie was Marvin Minsky, then director of the MIT Computer Science and Artificial Intelligence Laboratory (CSAIL) and one of the most influential AI researchers.

RELATED TOPICS
See also
THE GOLEM & MAGIC
page 14

ISAAC ASIMOV
page 22

FAMOUS 'INTELLIGENT' COMPUTERS
page 36

3-SECOND BIOGRAPHIES
ARTHUR C. CLARKE
1917–2008
British science-fiction author and science writer

MARVIN MINSKY
1927–2016
Pioneering American AI researcher, best known for *The Society Of Mind* (1986), a book about how the mind might be constructed

30-SECOND TEXT
Andreas Matthias

Will all AI machines like HAL conclude that humans are redundant and should be taken out of the way?

GREY WALTER'S TURTLES

the 30-second data

3-SECOND BYTE
'Turtles' was the nickname given to the pioneer *Machina Speculatrix* autonomous robots, conceived to help understand brain cells' interconnections and their impact on the emergence of complex behaviours.

3-MINUTE DEEP LEARNING
Today's robotics researchers are often influenced by Biology, Neurosciences and Psychology, and even Economics and Sociology. They believe that since nature has developed sophisticated animals, including humans, through thousands of years of evolution, it is a good source of inspiration for building better robots. Humans interact readily, so their systems provide a good basis for constructing fluid robotic devices.

The first two *Machina Speculatrix* turtles, built between 1948 and 1949 by the neurophysiologist William Grey Walter, were named Elmer (ELectro-MEchanical Robot) and Elsie (Electro-mechanical robot, Light Sensitive with Internal and External stability). They both had a front steering and driving wheel, driven by two independent motors, and two passive back wheels. One photoelectric cell and one mechanical switch were the robot sensors. The turtle nickname comes from the cover shell. The turtles' 'brain' was implemented by analogue electronics, composed of valves and relays. An ingenious connection of the sensors to the motors through the electronic brain enabled the robots to detect shell collisions and consequently move away from obstacles, as well as to follow or move away from a light source. These robot behaviours were complemented by a 'default' behaviour of exploration – until a light or an obstacle was found – and an exception behaviour that would lead the turtles to a battery-charging station when a low battery level was detected. Walter would claim some years later that his turtles could display animal-like behaviours such as self-recognition by tracking their own front light when watching it reflected on a mirror, and mutual recognition – they tracked each other by moving towards the other's lights.

RELATED TOPICS
See also
MACHINE LEARNING
page 40

EMBODIED AI & COGNITION
page 46

INTELLIGENCE AMPLIFICATION
page 106

MULTI-ROBOT SYSTEMS & ROBOT SWARMS
page 122

3-SECOND BIOGRAPHIES
NORBERT WIENER
1894–1964
American Professor of Mathematics who established the science of cybernetics

WILLIAM GREY WALTER
1910–77
American-born British neurophysiologist, who influenced well-known roboticists working on behaviour-based robotics

30-SECOND TEXT
Pedro U. Lima

Grey Walter became attached to his 'turtles', but the feeling was probably not mutual.

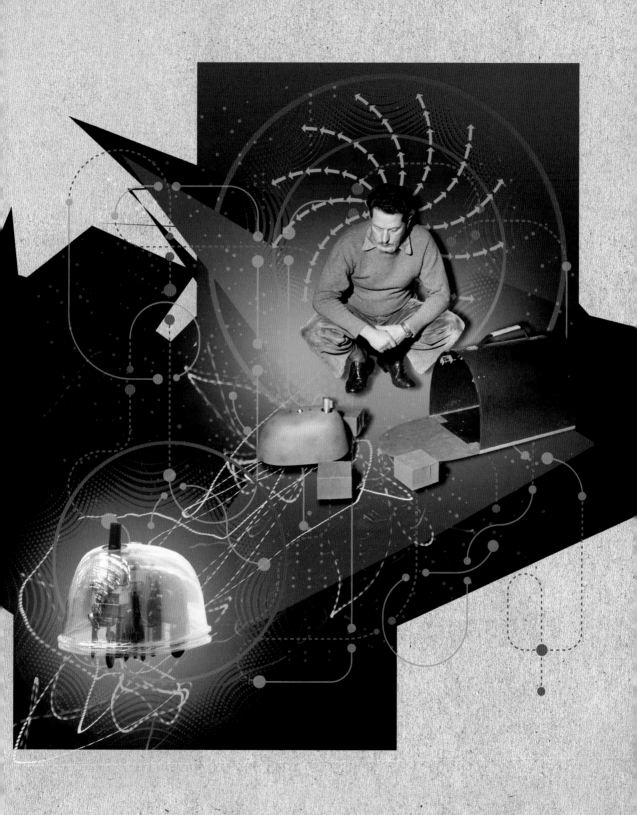

TERMINATOR & SKYNET

the 30-second data

3-SECOND BYTE
The Terminator introduced Skynet, an AI that wanted to exterminate all humans; Skynet has become the byword for villainous AI that poses an existential threat to humanity.

3-MINUTE DEEP LEARNING
Terminators are designed to mimic the appearance of humans, allowing them to infiltrate human resistance strongholds. This mimicking would be very difficult to achieve. In the 1970s Japanese roboticist Masahiro Mori found that not only are people good at noticing subtly non-human characteristics, but that also a human-looking robot that exhibited these subtle flaws caused a feeling of eerie revulsion. He called the effect the 'uncanny valley'.

The Terminator is a 1984 action film written by Gale Anne Hurd and James Cameron. Skynet, an artificial intelligence built for the US military, becomes self-aware. In response, humans panic and attempt to shut Skynet down. Skynet defends itself from this existential threat by attacking human civilization, launching nuclear weapons at Russia and relying on the Russian retaliation to destroy its enemies in the USA. With human civilization wrecked by the nuclear exchange, Skynet attempts to exterminate all remaining humans. The human resistance relies on a great leader, John Connor. Unable to kill him, Skynet sends a Terminator (a human-like cyborg assassin) back in time to kill Sarah Connor before John can be born. *The Terminator* taps into and reinforces some deep-seated Western fears about robotics and artificial intelligence. (Japanese culture, by contrast, tends to produce fiction in which robots are friends and helpers.) Firstly, Skynet is a human-like and yet far superior intelligence that becomes hostile to human beings the moment it becomes self-aware. Secondly, the Terminator is not self-aware, but physically resembles humans. Therefore the story utilizes two almost opposing fears: robots as emotionless unthinking aggressors – our dangerous inferiors – and computers as transcendent intelligences that have surpassed human intellect.

RELATED TOPICS
See also
MILITARY AI & ROBOTICS
page 66

IS THE INTERNET A HIVE MIND?
page 136

3-SECOND BIOGRAPHIES
DENNIS FELTHAM JONES
1917–81
British author whose book *Colossus* describes an intelligent super computer using nuclear weapons to attack humans

MASAHIRO MORI
1927–
Japanese robotics professor who developed the 'uncanny valley' hypothesis and started the Robocon robotics contest in 1981

30-SECOND TEXT
David Rickmann

Terminators tend to lack empathy and look a bit too hostile for a good conversation.

ARTIFICIAL INTELLIGENCE & COMPUTERS

ARTIFICIAL INTELLIGENCE
& COMPUTERS
GLOSSARY

algorithm A numbered plan of actions or formula for solving a problem or performing a specific task. While a computer program is generally viewed as the typical algorithmic system, a recipe in a cookery book is also a form of algorithm if it is followed rigorously.

artificial neural networks AI computing systems that are designed to work like a brain. They learn progressively by connecting inputs of information with outputs, through a filtering process that is partly autonomous, and via a collection of nodes called artificial neurons, in which each connection is a simplified version of a synapse. This model is responsible for the success of deep learning since 2011.

cryptography The science of making communication secure and readable only by those who have access to the appropriate technology. Cryptography involves codes, authentication, passwords and sometimes encryption (see below). In a highly digitized world, the protection of data and access to systems can be a matter of life and death.

cyborg A cybernetic organism, a hybrid being composed of flesh and mechatronic parts. A cyborg could be a bionic robot that has human features, or a human enhanced with electronic and mechanical parts.

encryption Related to cryptography, the process of encoding a message so that those who do not have authorized access cannot make sense of it. The primary purpose of encryption is confidentiality, and it is now available for many common computer programs and devices, such as apps, smartphones and financial systems.

language recognition A machine or program needs to be able to determine in which natural language a given set of data is formulated. Natural-language processing is an area of computer science and AI that works on the communication interfaces between machines and humans. This can be written or spoken language. Today, language recognition is one of the most advanced and successful fields of AI, and often involves deep learning networks.

minimax In game theory, minimax designates a decision rule that minimizes the possible loss for a worst-case scenario of maximum loss. If you have an opponent who is likely to win, minimax is a strategy to make them gain as little as possible, so that you don't lose too much. A similar term, maximin, is used in philosophy of justice to describe political and economic welfare groups that are beneficial for their least-advantaged members.

theory of mind Not a theory, but an idea that describes the fact that we attribute a mind to certain other beings, and no mind to others. Even if you think someone is an idiot, you believe that the person has beliefs, desires and knowledge, even if you disagree with them. We distinguish conscious beings from non-conscious beings. For robots and AI to interact with humans in a convincing manner, they will need to possess theory of mind.

THE TURING TEST

the 30-second data

The Turing Test, originally called the 'Imitation Game', is a test for computer intelligence: a judge exchanges text messages with two candidates: a real human being and a computer. If the judge cannot reliably distinguish the human from the computer, then we are justified in calling the computer 'intelligent'. Despite many criticisms, the Turing Test remains an influential idea in AI, and programmers are still competing in the attempt to develop similar tests (such as for the annual Loebner Prize). Critics of the Turing Test point out that imitating human behaviour in order to fool a judge is not the same as possessing intelligence. For example, a self-driving car undeniably does something intelligent when it drives around city streets, but it cannot be mistaken for a human in conversation, and would thus fail the Turing Test. There are also programs that are non-intelligent, but relatively good at chatting: ALICE (alicebot.org) is an example. ALICE uses a set of a few thousand pre-programmed sentence patterns. When it recognizes a trigger sentence typed in by the human operator, it will output a canned response. Although no real understanding of language is involved, ALICE won the Loebner Prize as 'most human-seeming' program three times.

3-SECOND BYTE
Since 1950, the Turing Test for machine intelligence has been the holy grail of AI accomplishment but its results are contested.

3-MINUTE DEEP LEARNING
John Searle's Chinese Room argument counters the Turing Test. Imagine a closed room with two slits: one for entering cards with questions in Chinese, and one for receiving answers in Chinese. Inside the room is a non-Chinese speaker, but he has a rule book that tells him which character to draw in reply to which question. He might fool outside observers that he speaks Chinese. Similarly, passing the Turing Test doesn't guarantee that a machine actually understands.

RELATED TOPICS
See also
ALAN TURING
page 38

CAN MACHINES THINK?
page 50

CHAT BOTS
page 92

3-SECOND BIOGRAPHIES
JOHN SEARLE
1932–
American philosopher; a critic of the Turing Test

HUGH LOEBNER
1942–2016
American sponsor of the Loebner Prize; contestants compete to develop a computer to pass the Turing Test

RICHARD S. WALLACE
1960–
American inventor of AIML, a programming language for chat bots

30-SECOND TEXT
Andreas Matthias

The Turing Test relies on the principle that most humans should be easy to imitate, because they are quite predictable

FAMOUS 'INTELLIGENT' COMPUTERS

the 30-second data

3-SECOND BYTE
Human intelligence and computer intelligence are still very different and incomparable. However, in competing with humans on specific tasks with precise rules, such as games, computers have the edge.

3-MINUTE DEEP LEARNING
Both Deep Blue and AlphaGo relied on an evaluation function: a mathematical model that assigns a value score to a move in a particular situation in the game. Deep Blue's commands were written in a programming language that software engineers could write, read and modify, whereas AlphaGo's operations were coded inside neural networks. In theory, a Deep Blue programmer could explain every move, whereas some AlphaGo winning moves caught everybody by surprise.

Deep Blue was a chess-playing IBM computer that challenged the then-reigning world champion Garry Kasparov twice, in 1996 and in 1997. The project had originally started in 1990 at Carnegie Mellon University. Deep Blue lost a match to Kasparov in 1996 but, after a series of hardware and software enhancements, it won in 1997. Watson is another IBM product, built to play the American TV game show *Jeopardy!* The project started in 2007 under the supervision of computer scientist David Ferrucci. In early 2011, Watson beat two human *Jeopardy!* champions, in a televised charity event. AlphaGo is a program that plays the ancient Chinese game of Go. It was conceived by computer scientist David Silver at DeepMind, the British AI company founded in 2010 and acquired by Google in 2014. In 2015 AlphaGo beat 5–0 the reigning European Go champion Fan Hui, the first time a machine had such a clear victory over a human player. AlphaGo's winning streak continued: in 2016 it beat Lee Sedol, widely considered the best player of the decade, and in 2017 it successfully played online against an array of top international players, collecting 60 victories in a row.

RELATED TOPICS
See also
MACHINE LEARNING
page 40

CAN MACHINES THINK?
page 50

CAN MACHINES HAVE
COMMON SENSE?
page 96

3-SECOND BIOGRAPHIES
FENG-HSIUNG HSU
1959–
Taiwanese computer scientist who designed Deep Blue and moved to Beijing, China to work for Microsoft in the early 2000s

DEMIS HASSABIS
1976–
British AI researcher, computer-game designer, and founder of DeepMind in 2010

30-SECOND TEXT
Mario Verdicchio

If an activity has a set of well-defined given rules, human ambiguity eventually loses against the cold efficiency of the computer.

23 June 1912
Born in Maida Vale,
London, England

1926
At the age of 13, Turing
attends Sherborne School
in Dorset

1931–34
Attends King's College,
University of Cambridge

1935
Age 22, elected fellow at
King's College

1936
Publishes 'On
Computable Numbers,
with an Application to the
Entscheidungsproblem
[Decision Problem]',
the paper in which he
proposed the Turing
Machine as a thought
experiment

1939
Starts work at Bletchley
Park on methods to
decode German wartime
communications

1950
In his paper, 'Computing
Machinery and
Intelligence', he describes
what later was called the
Turing Test for machine
intelligence

1952
Criminal proceedings
against Turing for 'gross
indecency' because of a
homosexual relationship

7 June 1954
Dies from cyanide
poisoning in Wilmslow,
Cheshire, England

ALAN TURING

Alan Turing was one of the most brilliant scientists of the twentieth century. Born in 1912 in London, he was the son of a civil servant working in British India. In school he already showed signs of his later genius, reading Einstein's work when he was only 16 years old.

In a short life of 41 years, he worked in mathematics, logic, cryptography and theoretical biology. He was also a passionate runner, who almost ended up on the 1948 British Olympic team. Although not a philosopher, two of his ideas have strongly influenced the philosophy of artificial intelligence: the Turing Machine and the Turing Test. The Turing Machine is a hypothetical simple computer that can read from and write onto an endless storage tape. With the help of this thought experiment, Turing analysed which types of problems can in principle be solved by a computer program. His analysis, in the form of the Church-Turing thesis, remains valid to this day. The Turing Test is a test for computer intelligence (see page 34), in which a computer is considered 'intelligent' if it can be mistaken for a human in a typed conversation.

During the Second World War, Turing worked at the Government Code and Cypher School at Bletchley Park, attempting to decrypt German military communications. The German military at this time used an encryption system based on a machine called Enigma, which encrypted messages through a series of rotating wheels. Turing helped to work out how Enigma worked, and together with his collaborators he built a machine to decrypt German messages. Due to various administrative hurdles, his group could not access enough resources to build more of these machines. In 1941, Turing wrote a letter directly to Prime Minister Winston Churchill, asking for more support, and Churchill gave a direct order that Turing's group should receive all the help that it required. In this way, Turing's group contributed significantly to the Allied victory in the war.

After the war, Turing worked on computers, but also on chemistry and biology at the University of Manchester. In 1952, Turing was prosecuted for homosexual acts, which was a criminal offence in the UK at that time. He died in 1954 from cyanide poisoning. It has never been clearly established whether his poisoning was an act of suicide or just an accident. Turing is considered today one of the 'fathers' of both computer science and artificial intelligence.

Andreas Matthias

MACHINE LEARNING

the 30-second data

3-SECOND BYTE
Machine Learning (ML)
refers to computer
algorithms (sets of rules)
that learn to classify and
predict data, rather
than being explicitly
programmed.

**3-MINUTE DEEP
LEARNING**
Reinforcement learning
algorithms were originally
inspired by psychology
studies, the most famous
example being the
attempt to understand
how rats learn their way to
cheese in a maze. When
the rat finds the maze for
the first time by moving
towards it from a nearby
maze cell, the animal is
rewarded. Simultaneously,
the algorithm increases
the weights of the
inter-neuron connections,
which reinforce the
probability of choosing
the same action when
in the same situation in
the future.

ML algorithms are trained to
build up an internal function to classify input
data. They use the resulting map to classify
new incoming data. An ML algorithm is able to
distinguish faces from different people after
being trained with images of those people's
faces, taken from a variety of viewpoints.
During training, the different images of the
same face are clustered in the same class,
labelled with that person's name – for example,
Fred. When a new image of Fred's face is
presented, the algorithm classifies it in the most
similar cluster – the one labelled as Fred. There
are three main types of ML algorithms. In
supervised learning, examples of previously
classified data samples are shown to the
algorithm, which adjusts the internal function
to classify new incoming data in similar classes.
In unsupervised learning, the computer finds
patterns directly from the unclassified data,
clustering them into classes with similar
characteristics. Reinforcement learning is a form
of supervised learning in which the algorithm is
not told the class of each training sample but
rather whether the classification is correct
(reward) or not (penalty).

RELATED TOPICS
See also
LANGUAGE RECOGNITION
& TRANSLATION
page 44

INTELLIGENCE AMPLIFICATION
page 106

GEOFFREY HINTON
page 128

3-SECOND BIOGRAPHIES
DONALD HEBB
1904–85
Canadian psychologist, best
known for his Hebbian theory,
which explains that as neural
connections are used more
frequently, they become
stronger and faster, and
learning occurs

PAUL WERBOS
1947–
American computer scientist
who in 1974 introduced a
method for training artificial
neural networks

30-SECOND TEXT
Pedro U. Lima

*Deep thinking used to
be an intuitive art. It is
now a way of projecting
probabilities into the
near future.*

VIDEO GAMES & AI

the 30-second data

3-SECOND BYTE
Video-game AI is typically concerned with making characters, opponents and environments more realistic, sometimes through generating an illusion of intelligent behaviour.

3-MINUTE DEEP LEARNING
Games that go beyond entertainment to address other purposes, such as research, education or physical therapy, are called serious games. These games offer controlled environments to test algorithms and to collect vast amounts of user data in small-scale models of the real world. Some studies on autonomous vehicles are conducted within a game environment to develop better navigation policies and build sophisticated AI techniques to improve the vehicle's perceptions of its surroundings.

While AI for academic games is mostly focused on developing self-learning systems, such as the AlphaGo, to enhance human performance, video-game AI is mostly concerned with improving the player's gaming experience. One aspect of this is programming convincing non-player characters (NPCs) whose behaviours are artificially generated rather than input by a human. However, it is often sufficient that NPCs behave in a barely intelligent way, without using sophisticated AI techniques. In the Pac-Man arcade game, the player navigates within a food-filled maze while running away from ghosts. Even though the movement of these ghosts seems random at first sight, their behaviours are strictly deterministic. Despite their simplicity, these behaviours make Pac-Man one of the most iconic games of our age. AI is often capable of beating human players in video games, due to multitasking and superhuman swiftness. However, humans are still better at games that require strategic creativity. To tackle this shortcoming, the AI can 'cheat'. Cheating in AI refers to providing artificial opponents with an advantage through extra tools, actions, resources or information that are normally unavailable for the human player. This kind of AI cheating is commonly used in modern games to set difficulty levels in gaming.

RELATED TOPICS
See also
FAMOUS 'INTELLIGENT' COMPUTERS
page 36

MACHINE LEARNING
page 40

TAMAGOTCHI
page 88

3-SECOND BIOGRAPHIES
CLAUDE SHANNON
1916–2001
American mathematician, electrical engineer and cryptographer whose paper 'Programming a Computer for Playing Chess' (1950) presents the minimax algorithm to let a computer play chess

TORU IWATANI
1955–
Japanese video-game designer, and the creator of Pac-Man, first released in Japan in May 1980

30-SECOND TEXT
Ayse Kucukylimaz

Will games remain friendly experiences if we become incapable of winning?

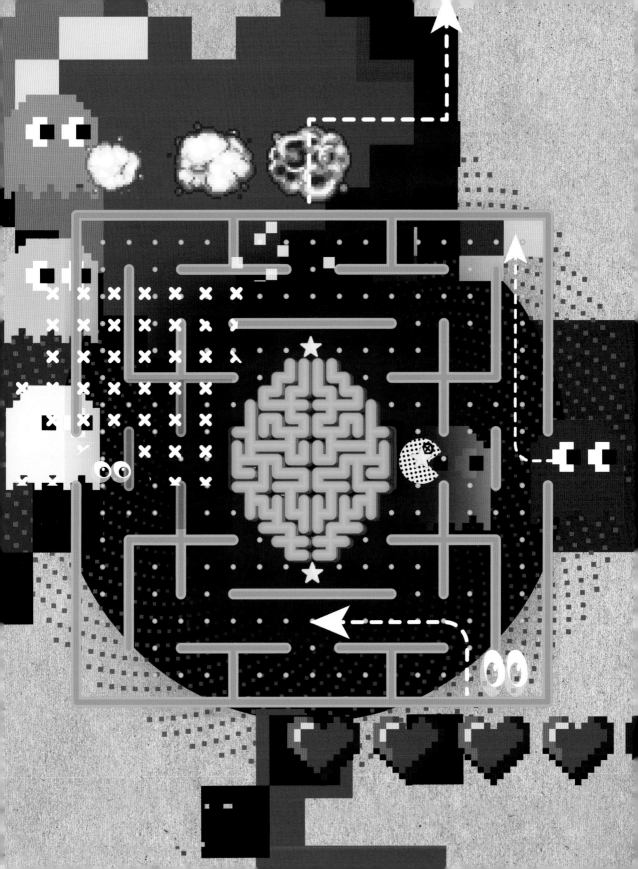

LANGUAGE RECOGNITION & TRANSLATION

the 30-second data

3-SECOND BYTE
Language is at the heart of human intelligence, and therefore at the centre of the AI race towards human-like levels of dialogue and understanding.

3-MINUTE DEEP LEARNING
Computers do not have yet a 'theory of mind' that would allow them to infer goals and beliefs, and take an interpretive decision based on a given personality and intimate history. They also lack 'common-sense reasoning'. Dialoguing humans are constantly making interpretations based on experience and intuition. In fact, for some philosophers, mind and language don't really differ; humans are constantly talking to themselves through an inner monologue.

'Our children and grandchildren will think it is completely natural to talk to machines that look at them and understand them', says Eric Horvitz, computer scientist at Microsoft's research laboratory. The idea of machine conversation dates back to philosophers René Descartes and Gottfried Wilhelm Leibniz, who proposed the idea of a universal symbolic language. In 1964, Joseph Weizenbaum, a professor at MIT, built the first chat bot program. Called ELIZA, it was programmed to act like a psychotherapist, asking questions to encourage conversation: 'What else comes to mind when you think about your mother?' Since the 2010s, researchers at companies such as Google, Facebook and Amazon, and at leading academic AI labs, are using machine learning to improve the quality of language understanding. From 2016, Google Neural Machine Translation system (GNMT) took advantage of deep neural networks to translate entire sentences. These networks use an opaque self-created intermediary grammar to translate between a pair of languages, thus rejuvenating Leibniz's dream. This 'interlingua' enables it to operate effectively, but it is not understandable by humans! In parallel, voice and speech recognition, via natural language processing (NLP), are approaching a human-like level of accuracy.

RELATED TOPICS
See also
CAN MACHINES THINK?
page 50

HOME ROBOTS & SMART HOMES
page 80

CHAT BOTS
page 92

3-SECOND BIOGRAPHIES
GOTTFRIED WILHELM LEIBNIZ
1646–1716
German polymath and philosopher who dreamt of a global 'concept language' he called *mathesis universalis*

JOSEPH WEIZENBAUM
1923–2008
German-American computer scientist; from 1964 to 1966 he developed a natural language program that simulated a conversation with a psychotherapist

30-SECOND TEXT
Luis de Miranda

There was a time when speaking to objects was considered pure madness.

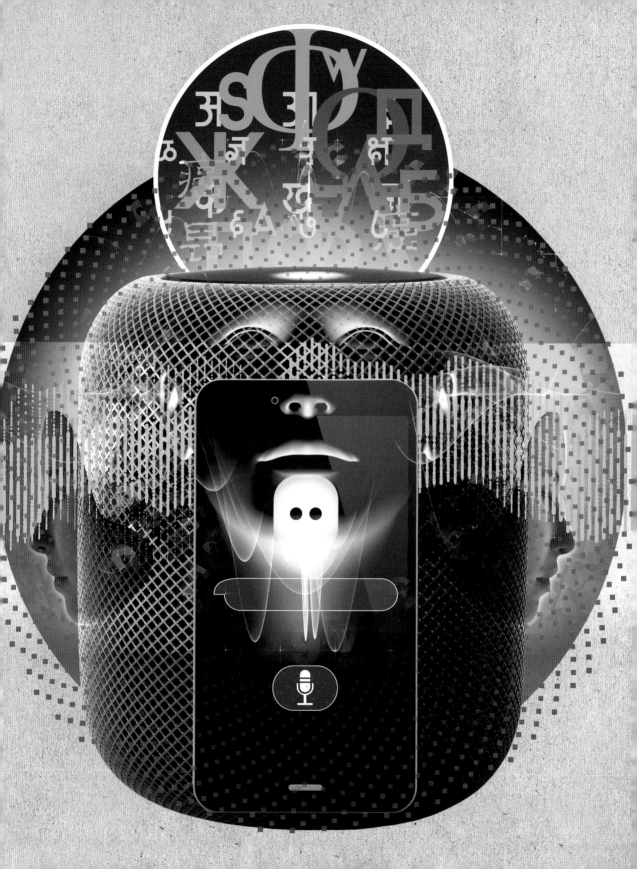

EMBODIED AI
& COGNITION

the 30-second data

Does cognition require a body?

One can calculate or play chess without a body. But cognition is not only abstract thought. It includes awareness of one's surroundings, the ability to react to an environment and to be purposefully active in it. Much of our common sense comes from having a body. Many mental concepts, like 'red' or 'salty' can only be understood by pointing to a bodily experience. Remembering can be performed more efficiently with the aid of external physical resources. And we have nerve cells, so-called 'mirror neurons', which fire when we perform an action and when we see others perform it. This forms the basis for empathy, for 'feeling with' someone else, which is linked to moral behaviour. Some intelligent behaviours are physical; after a lot of practice, walking, opening a door, and even writing become automatic behaviours. Philosopher Hubert Dreyfus sees this 'skilful coping' as what human intelligence is really about. Moreover, many cognitive abilities of the individual are developed within a social body, such as a school or other institution. Most of our knowledge comes from our culture, which makes it impossible to separate individual cognition from culturally determined behaviour. Human-like cognition appears to require a body that at least roughly resembles ours, and a comparable socialization process.

RELATED TOPICS
See also
RODNEY BROOKS
page 64

HUBERT DREYFUS
page 148

THE MERGING OF ROBOTS,
AI & HUMANS?
page 150

3-SECOND BIOGRAPHIES
MAURICE MERLEAU-PONTY
1908–61
French philosopher who emphasized the role of the body in understanding the world

BRUNO LATOUR
1947–
French philosopher who saw humans and machines fusing into 'collective' agents

30-SECOND TEXT
Andreas Matthias

If having a body (or belonging to social bodies) defines who we are and how we think, then our robots might think and act differently from us.

3-SECOND BYTE
As long as computers are grey boxes without a body, will they ever be able to really understand us – and will we be able to understand them?

3-MINUTE DEEP LEARNING
We already share our bodies with intelligent machines. A mobile phone radically enhances the capabilities of its user, making the person into a cyborg, a life form that is improved by incorporating technology. The phone is embodied through its user, sharing the user's ability to move around. Technologies often carry social biases and values, turning us into social cyborgs. As we incorporate more technologies into our bodies, the limits between individual, machine and society become increasingly unclear.

ART & AI

the 30-second data

3-SECOND BYTE
Art is considered a specifically human form of expression and is therefore a challenge for AI technology, which aims to reproduce and enhance all human endeavours.

3-MINUTE DEEP LEARNING
Exploiting Google's image database and its own recognition and matching capabilities, the company's Deep Dream project came up with a new psychedelic visual style. Arguably, this is not truly creative since Deep Dream's works are based on existing images, but human artists also base work on images they have seen. The real limitation is the lack of symbolic meaning that a human artist can convey, thanks to a shared cultural context. How to code context into computers still eludes AI researchers.

Computation and art have a long history together: perspective in drawing has been coded in the form of mathematical rules since the fifteenth century. More recently, it took less than a decade after the introduction of digital computers in the 1950s for Computer Art pioneers such as Frieder Nake, Georg Nees and Michael Noll to exploit computer programs to create abstract art. The strongly geometric appearance of these works revealed the intrinsically numerical nature of the techniques behind them. In contrast, Harold Cohen's computer program AARON – in continuous development from the 1970s to the early 2000s – produced visual works that were more representational, and were characterized by a distinct, recognizable style that could have been associated with a human painter. However, hardly anybody ever talked about the computer as an artist. In these endeavours, computers were mere instruments in the expert hands of human programmers: creativity was still a strictly human characteristic. This situation changed in the 2010s, when technological development made neural network-based techniques possible. Neural networks can be trained to recognize, match and reuse any numerical pattern. Neural networks are now used in several research labs for experiments in music composition and automatic storytelling.

RELATED TOPICS
See also
FAMOUS 'INTELLIGENT' COMPUTERS
page 36

MACHINE LEARNING
page 40

CAN MACHINES THINK?
page 50

CAN MACHINES HAVE COMMON SENSE?
page 96

3-SECOND BIOGRAPHIES
HAROLD COHEN
1928–2016
British-born creator of AARON, which could produce art 'autonomously'

FRIEDER NAKE
1938–
German computer art pioneer who first produced digital art in 1963

30-SECOND TEXT
Mario Verdicchio

Art without artists is like society without consciousness.

CAN MACHINES THINK?

the 30-second data

'Can machines think?' What is this question really asking, and what is thinking? When we talk about thinking, formal thought first comes to mind, such as mathematics, logic or chess. Machines can do all these things, often better than humans can. But can they go beyond formal thought? Alan Turing had the idea that if a machine could fool a human into believing that it was also human during a typed conversation, then the machine would be able to think. But is this true? Chess or Go-playing computers don't talk and neither do self-driving cars. Some humans don't talk, either because they are mute or they just don't want to. Can we conclude that people don't think? Obviously not. And what about emotions? Is having feelings a necessary part of thinking? Is having emotions perhaps even sufficient to say that someone is thinking? But then, dogs appear to have emotions. Does this mean that they think? It seems that there is a spectrum of behaviour from not thinking to highly developed thinking, from that of an insect up to that of a mathematician. Therefore, whether machines can think cannot be simply answered by 'yes' or 'no'.

3-SECOND BYTE
The question whether artefacts can think has intrigued philosophers and engineers since ancient times – the problem is that we still don't know what the question itself means.

3-MINUTE DEEP LEARNING
Influential German philosopher Martin Heidegger argues that our functionalist, technological understanding of intelligence misses the embodied richness of human existence. To be, to think and to share experiences with the world are indistinguishable from each other. By living in the world, we realize our being through our choices and actions, but a machine cannot do that. A machine is always a tool, created for the sake of something else, rather than for its own sake.

RELATED TOPICS
See also
THE TURING TEST
page 34

CAN MACHINES HAVE
COMMON SENSE?
page 96

HUBERT DREYFUS
page 148

3-SECOND BIOGRAPHIES
MARTIN HEIDEGGER
1889–1976
German philosopher, now recognized as one of the most influential philosophers of the twentieth century

MARVIN MINSKY
1927–2016
American cognitive scientist who explored thinking processes in his book
The Society of Mind

30-SECOND TEXT
Andreas Matthias

What if deep thinking was about the overcoming of all automatic thoughts?

INDUSTRIAL, CORPORATE & INSTITUTIONAL AI & ROBOTICS

INDUSTRIAL, CORPORATE & INSTITUTIONAL AI & ROBOTICS
GLOSSARY

aerial robotics Phrase coined by US engineer Robert Michelson in 1990. Flight robots are also often called drones. Their purpose is to perform unmanned missions in which they are self-navigating and able to interact with their environment and objects on the ground. Typical applications are military related, such as reconnaissance or demining, but aerial robots are also used for product delivery, photography, agriculture, smuggling, law enforcement and surveillance, disaster relief and scientific research.

cobot Portmanteau word for 'collaborative robot' proposed in 1995 by the research labs of General Motors. A cobot is designed to interact with humans in a shared workplace, supporting human operators in specific tasks that demand precision or force. Usually a cobot can adjust to its environment and detect abnormal activity via a system of sensors. Cobots are meant to be safer than robots, easily programmable by non-specialists, and flexible in the tasks they perform.

cyber The term comes from cybernetics, the study of control and communication in systems involving machines and humans. It often designates anything digital and connected to the internet, such as cyberspace, cyberbullying or cybercrime – crime that involves the use of a computer and a digital network.

hand guiding Technology connecting a human hand with a robotic hand that is usually intended to imitate it. The human user generally faces a camera that tracks the position of the human hand. The position of the hand is continuously updated, and the robot mimics the movement. A typical task for an industrial robotic hand is to pick up something in one location and move it to another.

heuristic Based on the Ancient Greek root meaning 'to discover', a heuristic approach is the creative use of any device, method or rule of thumb that will facilitate a discovery, even if this method does not seem directly correlated to the orthodox way things are usually done. In engineering, heuristics are experience-based methods used to reduce the need for calculations.

power and force limiting The use of cobots implies that incidental contact between a moving robot and a human body can occur. Power and force limiting (PFL) designates a technology that is meant to avoid injury from such a contact, or even to avoid it. This involves a minimal form of sensory perception so that the cobot can adjust to its environment. Whenever the robot initiates contact, it will represent a certain force, F, the power of which must be autonomously limited in an environment that is constantly changing.

proprioceptive senses Proprioception is the human bodily sense of itself as a whole and can be conscious or unconscious. Proprioceptive senses primarily include the sense of position and the sense of effort. In robotics, the challenge is to provide a machine with a minimal sense of its own integrity so that it can avoid self-damage.

safety-monitored stop Some robots are designed to stop operation when a human being enters the workspace. Safety-rated monitored stop allows the operator to interact with the robot when it has stopped, and operation automatically resumes when the person leaves the collaborative workspace. As the presence of cobots in workplaces increases, safety will be a key concern for manufacturers.

sense-and-avoid technologies Technologies concerning unmanned aerial vehicles, their purpose is to provide an aerial robot with the capacity to feel potential obstacles, such as birds or planes, and avoid them autonomously. Sensors collect and record data along the flying path, which is continuously analysed by a collision-avoidance program.

speed and separation monitoring Technology aimed at controlling the separation distance between a human and a robot. This distance must remain at all times protective, even when the human and the cobot perform tasks concurrently. The machine has to have safety-rated monitored speed and stop functions. An important property of the sensors includes the capacity to self-evaluate under which conditions the system should return to its normal mode of functioning.

universal automation Concept patented by US inventor George C. Devol in 1954, which he shortened to Unimate. It described a robotic arm that was programmable and 'teachable'. Today, the concept of artificial general intelligence includes the idea that an AI could perform a variety of tasks with great flexibility.

UNIMATE, THE FIRST INDUSTRIAL ROBOT

the 30-second data

Humans are born with a full body

but robots first existed as mere arms: automated handling devices. Unimate, the first industrial robotic arm, was created by US polymath inventor George Devol. It was installed in 1961 at the New Jersey General Motors plant, efficiently lifting and welding hot metal parts with its hydraulically driven joints. Initially, it could operate 150 hours without technical failure (and later up to 400 hours). Along with entrepreneur Joseph Engelberger, Devol created the company Unimation, a portmanteau for universal automation – the idea of flexible robots fulfilling many tasks and evolving over time. Rather than transistors, the machine was originally controlled by vacuum tubes used as digital switches. In the 1960s, human labour was abundant and US industry in a dominant position, so many regarded automation as unnecessary. But a team of Unimates could help produce up to 110 cars per hour! Soon, international car manufacturers (Fiat, Chrysler, Volvo and Ford) purchased their own version of Unimate, but it was the labour shortage in Japan and a collaboration with Kawasaki that really made Unimation's fortune. In the 1980s, the 'Isolation Syndrome of Automation', a form of depression, appeared in Japanese humans who worked with robots all day long.

3-SECOND BYTE
A robotic arm implemented in the USA in the 1960s, Unimate was the most important automated device of the twentieth century, revolutionizing the car industry and the workplace.

3-MINUTE DEEP LEARNING
Collaboration with robots in the workplace is common today; many jobs are being automated. Collaborative robots are called cobots when the fencing that used to separate them from humans is removed. Cobot technology includes a safety-monitored stop (stops when the safety zone is violated), hand guiding (a human hand holds the arm), speed and separation monitoring (a laser tracks the position of the human worker) and power and force limiting (the robot can feel abnormal forces in its path and adjust).

RELATED TOPICS
See also
HUMAN-ROBOT INTERACTION
page 78

INDUSTRY 4.0
page 142

THE MERGING OF ROBOTS, AI & HUMANS?
page 150

3-SECOND BIOGRAPHIES
GEORGE DEVOL
1912–2011
American inventor; as well as the Unimate, he was involved in inventing microwave ovens, magnetic-recording devices, and automatic opening doors

JOSEPH ENGELBERGER
1925–2015
American entrepreneur and engineer. The first major entrepreneur in the robotic industry, he was Devol's partner in 1956 when they created Unimation, the world's first robotics company

30-SECOND TEXT
Luis de Miranda

Objects become universal makers as humans become mere users.

EXPERT & SECURITY SYSTEMS

the 30-second data

RELATED TOPICS
See also
CAN MACHINES THINK?
page 50

COLLECTIVE AI & CLOUD
ROBOTICS
page 124

HUBERT DREYFUS
page 148

An expert system is a heuristic software: it helps humans find solutions by interpreting input, predicting results and suggesting alternative options. These systems ask questions like 'what could happen next?' or 'why did this happen?' If you wish to create a car or robot capable of discovering its own faults – and even self-repairing them – you need an AI expert system. In hospitals, expert systems are used for diagnosis. Increasingly, companies are implementing them for security, such as video surveillance, facial recognition and cybersecurity to avoid damage to a computer network. Advances in machine learning have made security systems easier to train and more effective. A group of researchers at MIT's Computer Science and Artificial Intelligence Laboratory (CSAIL) recently built a system that could detect 85 per cent of cyberattacks. A more controversial use of expert systems is biometrics, which at the moment are mostly used for identification and controlling access to systems, and tracking people under surveillance. Some say that if applied to identify psychological traits or tendencies based, for example, on the shape of a person's face, or life habits or genetic data (at border controls, the workplace or on mobile phones), biometrics could lead to a form of cyber-fascist political regime that would control our existence.

30-SECOND TEXT
Luis de Miranda

Will we become overprotected? Protected from what we have not yet done and will never experience.

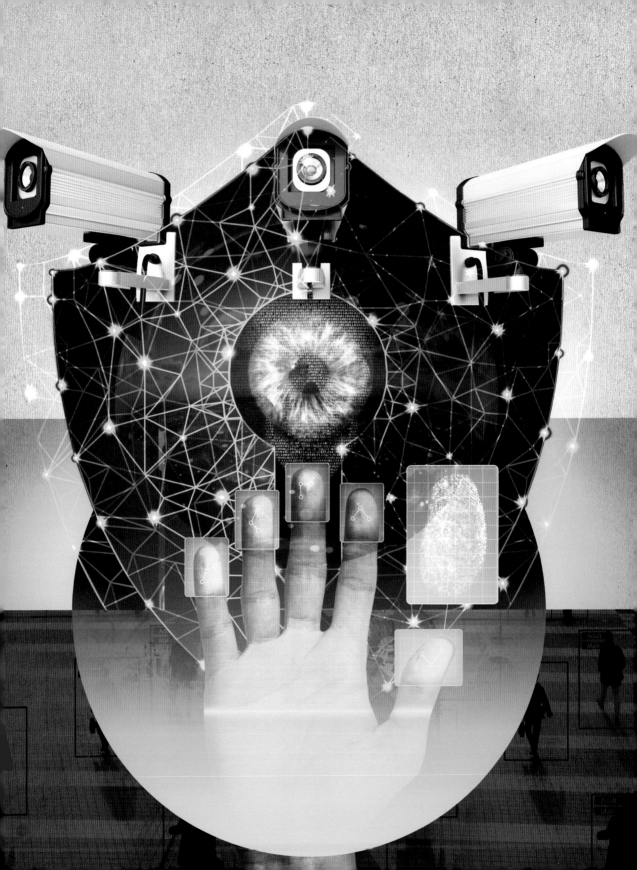

AI IN BANKING & FINANCE

the 30-second data

Most banks are investing millions in artificial intelligence in order to replace their human workforce, and in the hope of improving the efficiency of their procedures. Some repetitive tasks, such as fund transfers, are now fully automated. Some work that previously took thousands of hours now takes seconds, for example, the analysis of legal documents to extract important clauses. Chat bots or 'conversational interfaces' are also increasingly used as virtual assistants to communicate with human users and provide account information or help with online banking. More sophisticated 'robo-advisors', such as Erica from Bank of America, offer a form of financial guidance. Real-time machine-based learning is used to detect patterns of fraud via analysis of users' big data. In stock trading, machine learning makes it easier to identify patterns that might not otherwise be detected by the human eye. More recently, banks have been investing in deep learning. By analysing unstructured data such as financial information on news sites or across social media, they aim to generate faster and smarter trading. Digital cryptocurrencies, such as Bitcoin, are confidential electronic payment systems that could potentially replace the need for central banks and many human jobs in banking and finance.

RELATED TOPICS
See also
CHAT BOTS
page 92

IS THE INTERNET A HIVE MIND?
page 136

INDUSTRY 4.0
page 142

3-SECOND BYTE
Since the first automatic teller machines (ATMs) in 1967, computer software and hardware have slowly replaced humans in banking and finance.

3-MINUTE DEEP LEARNING
Experts attribute the acuteness of the 2008 global financial crisis partly to the fact that opaque computer programs entered into a destructive loop that snowballed across the financial system. Financial systems are all interconnected and often use similar equations and software, which adds to the potential dangers. The need for human oversight puts ethical and legal limits on using AI for finance. AI can help with black-and-white decisions, but the shades of grey will probably still be decided by humans for a while.

3-SECOND BIOGRAPHIES
JOHN SHEPHERD-BARRON
1925–2010
British inventor who led the team that installed the first cash machine or ATM (Automated Teller Machine) in 1967

ANGUS DEATON
1945–
British-American economist who won the Nobel Prize for his work on health and wealth, believes that robots will cause a major global financial crisis worse than globalization

30-SECOND TEXT
Luis de Miranda

New currencies and new automated transactions might reinforce capitalism or accelerate its fall.

MEDICAL AI & ROBOT-ASSISTED SURGERY

the 30-second data

Minimally invasive surgery (MIS) was revolutionary when it first took off in the 1990s, reducing recovery time and causing less pain and fewer complications. By the time MIS became popular the following decade, researchers were focusing on making the process robot-assisted. Today, MIS can be done using robotic helpers. In robot-assisted surgery, the surgeon is provided with a magnified 3D view of the surgical site, and controls the robot through a tele-operation interface. The robot's purpose is to increase the precision, flexibility and control of the surgeon. The cost of a commercial surgical robot can reach £1.7 million, plus hundreds of thousands of pounds in annual servicing costs. The cost is slowing down the global uptake of robotic technology. Also, after almost 17 years of use, evidence to prove that robotic technology provides significant benefits is still lacking. As of 2018, unmanned surgery had only been tried out on pigs. Even though the results are promising, from a social and ethical point of view, the idea of using fully autonomous robotics systems for surgery is controversial. Problems with the workspace, dexterity of the machine or errors in the robot's sensory system could potentially prove fatal for the patient.

Will we learn to be suspicious of any human involved in medical processes?

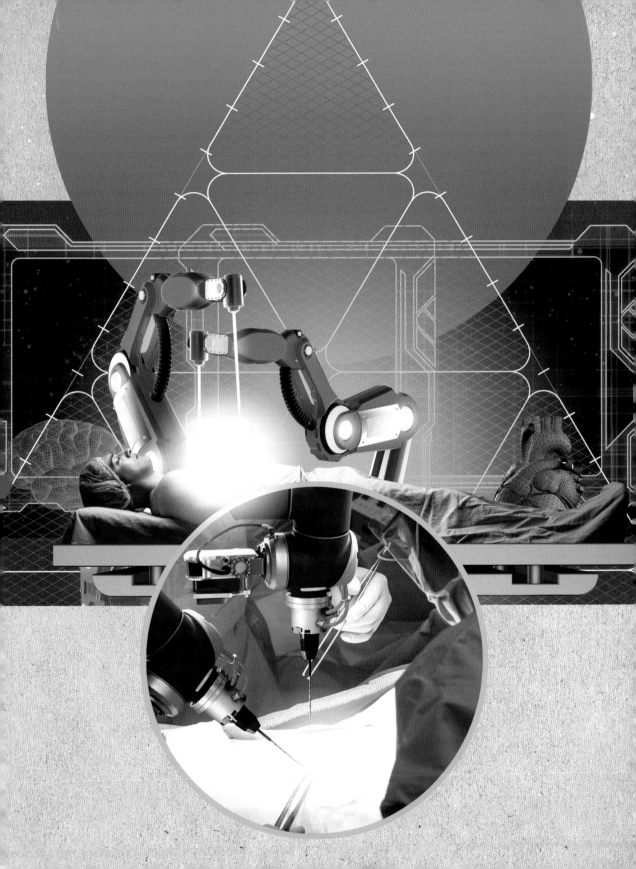

30 December 1954
Born in Adelaide,
Australia

1978
Master's degree in pure
mathematics (Flinders
University)

1981
PhD in computer science
(Stanford University)

1984
Joins the Mobile Robotics
Laboratory at the
Massachusetts Institute
of Technology

1990
Publishes the paper
'Elephants Don't Play
Chess'

1990
Co-founds the company
iRobot

1997
Director of the MIT
Artificial Intelligence
Research Laboratory

1999
Publishes *Cambrian
Intelligence: The Early
History of the New AI*

2008
Leaves MIT to co-found
the company Heartland
Robotics (later Rethink
Robotics)

2012
Introduces the industrial
robot Baxter

2015
Introduces the
collaborative robot
Sawyer

RODNEY BROOKS

'Over time there's been a realization that vision, sound-processing, and early language may be the keys to how our brain is organized', stated Rodney Brooks in his seminal paper, 'Elephants Don't Play Chess', in 1990. From then on, Brooks revolutionized the field of robotics by arguing that, in order for robots to accomplish everyday tasks in an environment shared by humans, their cognitive abilities needed to be based on sensory-motor action, stimulus behaviour, and proprioceptive senses rather than on top-down logical models. Proprioception is our body's ability to perceive our environment with our entire body and senses. In other words, to build better robots, the animal model was better than the computational model. At that time, many specialists in the field claimed Brooks' career was over before it started, but in fact he opened up a new field called the 'actionist approach to robotics'. According to the 'Nouvelle AI' approach, as it was called in the 1980s, robots were thought of as situated and embodied beings. The situated approach builds agents designed to interact with their environment, considering their body and simple sensing first. Insects were the primary model for Brooks, because they are able to perform all sorts of complicated tasks without much analytical thinking. With his MIT team, Brooks created Allen, a robot that resembled a footstool on wheels and used sonar distance perception and odometric (from the Ancient Greek meaning 'measuring the route') motion sensors to estimate change in position over time.

Rodney Allen Brooks was born in South Australia and became passionate about the first personal computers when he was a student in mathematics. Since his PhD on computer vision at Stanford University, awarded in 1981, he has led a double life as a respected academic, mostly at the Massachusetts Institute of Technology (MIT), and successful entrepreneur. In 1990, he co-founded the company iRobot with former students Colin Angle and Helen Greiner, a brand that became famous for selling more than 10 million of its autonomous robotic vacuum cleaners, Roomba. In 2008, he co-founded Heartland Robotics with Ann Whittaker in Boston, with the goal to create collaborative industrial robots (cobots). Later called Rethink Robotics, the company introduced Baxter and Sawyer, cobots working on an intuitive software that allows non-technical staff to train and interact with them. Rodney Brooks is also a philosopher of technology who often writes about AI and robotics issues in his popular blog.

Luis de Miranda

MILITARY AI & ROBOTICS

the 30-second data

3-SECOND BYTE
Military AI systems are used to supply strategic operational support, tactical support and to attack one another, but could also potentially attack human targets autonomously.

3-MINUTE DEEP LEARNING
In 2003, the United States deployed Patriot missiles in the Iraq War. One of the earliest US 'lethal autonomy' missile systems, it was designed to use lethal force with minimal human oversight. This was deemed necessary to allow it to track and fire on missiles within a very short engagement time. However, an over-reliance on this autonomy resulted in the system destroying two friendly aircraft, killing their crew. This may be the first instance of a lethal autonomous system killing humans.

In war any advantage may mean the difference between victory and defeat. Armies have always sought technological advantage: iron over stone, tanks over cavalry. Today, there is huge investment in AI and robotics. Military AI systems assist with strategy and data gathering. For example, AI systems operated by the US National Geospatial-Intelligence Agency have enhanced battlefield intelligence by analysing unstructured satellite data and producing a coherent story for human agents. AI systems also operate on a tactical level by autonomously monitoring the battlefield and tracking friendly and enemy positions. They can even manage communication systems, filtering available data and routing it to where it is needed. Humans decide how to act upon the data, so these systems are not strictly autonomous. However, the autonomy of such systems is increasing. The US military is developing AI systems capable of disrupting enemy AI, while defending itself from similar attacks. The 2016 DARPA Grand Cyber Challenge had seven systems competing to disable each other, and defend themselves, without any human intervention. The next step is to remove humans from the loop entirely. In 2006, the first prototype of an autonomous gun platform was deployed by South Korea, capable of targeting without human supervision.

RELATED TOPICS
See also
TERMINATOR & SKYNET
page 28

CAN MACHINES HAVE COMMON SENSE?
page 96

BIONIC PROSTHETICS & EXOSKELETONS
page 104

3-SECOND BIOGRAPHIES
ZHUGE LIANG
181–234
Chancellor of the State of Shu Han, China in the third century, he is credited with the invention of the land mine

HUGO GERNSBACK
1884–1967
Luxembourgish-American pioneer of science fiction who wrote an article in October 1918 entitled 'Automatic Soldier', possibly one of the first ideas for automated warfare

30-SECOND TEXT
Lisa McNulty

Is there a form of killing that is more acceptable than another?

AERIAL ROBOTICS

the 30-second data

Unmanned aerial vehicles (UAV) are systems capable of sustained flight with no direct human control, and able to perform a specific task. Although related machines were flying as early as the Second World War, the drone boom really started in Japan, with the development of reliable Yamaha helicopters in the 1980s. Today, most operational aerial robots have fixed wings. These drones are used for remote sensing (spotting or mapping, in agriculture or geology), disaster response (flood monitoring, wildfire management), surveillance (by law enforcers, or maritime or border patrols), search and rescue missions, cargo transportation, communications, image acquisition (filming from new angles) and warfare (bomb dropping or spying). Path planning, object detection, recognition and mission management often require distant human operators, who remain in contact with the flying machine. There are few differences between these remotely piloted vehicles (RPV) and traditional manned aircraft, except that the pilot is on the ground and does not 'feel' the mission. Researchers are now developing AI-based navigation technologies that allow aerial robots to operate with minimal human control, and even to handle situations that cannot be detected at a distance.

3-SECOND BYTE
Robots that fly, often called drones, are used both for military and civilian purposes: surveillance, targeting, sensing, disaster response, image acquisition and delivery of goods.

3-MINUTE DEEP LEARNING
Aerial robots are the theme of much discussion in regulatory agencies, which must find ways to insert them into airspace occupied by other traffic, such as traditional aircraft. This is connected with the current development of sense-and-avoid autonomous technologies (to detect an unpredicted obstacle such as a bird or a plane and prevent impact) and the ability for aerial robots to communicate with other traffic and the ground-control infrastructure to avoid collisions.

RELATED TOPICS
See also
MILITARY AI & ROBOTICS
page 66

SPACE EXPLORATION
ROBOTICS & AI
page 70

MULTI-ROBOT SYSTEMS
& ROBOT SWARMS
page 122

3-SECOND BIOGRAPHIES
REGINALD DENNY
1891–1967
British actor, aviator and UAV pioneer, he worked with engineer Walter Righter to develop the large-scale manufacture of radio-controlled UAVs

ROBERT C. MICHELSON
1951–
US engineer, academic and inventor, sometimes called the 'father of drones', who invented the entomopter, an aerial robot

30-SECOND TEXT
Luis de Miranda

When it comes to drones, is human control an obstacle on the path to total control?

SPACE EXPLORATION ROBOTICS & AI

the 30-second data

RELATED TOPICS
See also
HAL 9000 IN *2001: A SPACE ODYSSEY*
page 24

MACHINE LEARNING
page 40

3-SECOND BYTE
AI systems allow unmanned spacecraft to undertake more sophisticated missions and to operate at a greater distance from the Earth than would be possible with manned spacecraft or merely automated systems.

3-MINUTE DEEP LEARNING
A problem with space exploration is that there is so much data to churn through. For example, the Large Synoptic Survey Telescope (LSST) in Chile is set to take over 200,000 pictures every year. Reviewing all the data for interesting discoveries would be difficult for humans, but it is an ideal task for an AI system. AI researchers hope to use similar systems to analyse the enormous quantities of data from space observations around the world.

When we send humans to space we must provide air, food, water and a way to bring them back. Unmanned spacecraft escape these complications but, given the vastness of space, it is not possible to maintain direct control of them. Receiving and responding to a signal from a spacecraft orbiting Jupiter can take as long as 106 minutes. To combat this problem, many space probes have been automated, usually with a set of pre-planned instructions to follow rather than any true decision-making capacity. The first AI system to control a spacecraft independently was the Remote Agent Executive on board the Deep Space One probe, launched in 1998. Since then, autonomous systems have helped spacecraft operate at a greater distance from Earth and carry out more sophisticated missions. In 2015, scientists upgraded the Mars rover Curiosity with an AI system to allow it to choose targets for its geologic laser to sample, freeing up significant bandwidth on the Deep Space Network for additional mission planning. The 2020 Mars Rover is designed to have even greater autonomy and be able to navigate terrain independently. Similarly, the Europa Clipper will autonomously identify and analyse short-term events such as ice plumes. As AI improves, NASA engineers are even considering probes that will dive beneath the ice on Europa.

3-SECOND BIOGRAPHIES
FREEMAN DYSON
1923–
An English-born physicist who proposed a semi-biological, self-replicating probe, nicknamed the Astrochicken, which could explore space more efficiently than manned craft

MARGARET HAMILTON
1936–
American computer scientist who led the development of software for the Apollo moon landings and invented the term 'software engineer'

30-SECOND TEXT
David Rickmann

If aliens do not exist, then humans will create them.

AI IN THE ENTERTAINMENT BUSINESS

the 30-second data

3-SECOND BYTE
Part of daily life in the form of movies, music and video games, online entertainment is tailored for customers, who receive personalized recommendations identified with AI techniques.

3-MINUTE DEEP LEARNING
The broad adoption of recommender systems that filter results has been raising ethical concerns. For example, filtered recommendations may not inform customers about the variety of available products. In the context of news providers, this phenomenon is called the Filter Bubble: it isolates a customer inside a bubble, containing only the information selected for them, and shielding them from diverse sources. This may dramatically reduce customers' options.

Online entertainment providers use AI, in the form of recommender systems, to predict customers' preferences for movies and music and recommend to them what they are most likely to enjoy. A recommender system can be seen as a 'silent' smart assistant that works behind the scene: it knows, in theory, what each customer tends to like and helps them in navigating and selecting products from the vast catalogue the online entertainment provider offers. For the system to provide accurate recommendations, it requires information about each customer, such as demographic data, the products they have selected in the past and how they rated them. This information is then used by an algorithm to predict customers' preferences. Numerous scientific researchers have focused on the improvement of prediction techniques in recent years, and have designed sophisticated algorithms for recommender systems. The two basic approaches used in the design of such algorithms are based on the concept of similarity: items recommended to a customer may be similar to items she already liked or bought, or they can be items liked or bought by other customers the system thinks are similar to the first person.

RELATED TOPICS
See also
VIDEO GAMES & AI
page 42

SMART ASSISTANTS & 'LITERATE' MACHINES
page 86

CAN MACHINES HAVE COMMON SENSE?
page 96

30-SECOND TEXT
Sofia Ceppi

How can we deeply enjoy something that we are programmed to enjoy?

SOCIAL ROBOTICS

SOCIAL ROBOTICS
GLOSSARY

3D printing Technology enabling the creation of a physical object directly from a digital design, through the melting and shaping of a plastic material within a printing device. Since 2009, desktop 3D printers have become increasingly available. Doctors use them to manufacture tailored prosthetics, architects to create scale models and students to design prototypes.

anthropomorphic robot In robotics design, there are usually two schools of thought: one is to have robots to perform a specific task. The second is to try to engineer the complexity and agility of human biological structure. To design a robotic hand, some engineers believe that if we want the robot to evolve to be flexible and carry out many tasks, it has to be as human-like as possible.

cloud network Cloud computing relies on sharing of resources and storage: data on a cloud network is remotely maintained, managed, backed up, and always accessible and modifiable. Third-party cloud providers enable corporations or institutions to focus on their core practice instead of spending time and money on computer infrastructure and maintenance.

cyborg A cybernetic organism, a hybrid being composed of flesh and mechatronic parts. A cyborg could be a bionic robot that has human features, or a human enhanced with electronic and mechanical parts.

Defense Advanced Research Project Agency (DARPA) A national agency connected to the US Department of Defense that invests in long-term research perceived as beneficial for national security. Outcomes have included precision weapons and stealth technology, but also breakthroughs in civilian technologies such as the internet, automated voice recognition and language translation.

empirical analysis Empiricism is the idea that knowledge of the world around us comes primarily from experience. In machine learning and AI technologies, empirical data analysis is a data-driven approach that is supposed to be free from strong initial assumptions. Some people believe data-driven science and technology form a new paradigm in which operational patterns and theories will be elaborated by AI systems out of big data, without the help of a human theorist. Empirical analysis also describes the capacity for a robot interacting with a human environment to read its current properties in real time via sensors.

humanoid A less gendered term for android. 'Humanoid' can also describe a robot that has some kind of human feature, without being as human-like as an android or a gynoid. Most engineers believe that a humanoid robot is more easily accepted by humans. The same reasoning does not seem to apply to AI research: most AI systems today are expected to process information better than humans, and in ways that do not have to mimic our reasoning.

inertial measurement sensor An electronic device that senses and measures a body's force and movement, using sensors such as accelerometers and gyroscopes. The technology is used for aircraft vehicles and aerial robots. An accelerometer senses and measures acceleration, while a gyroscope measures changes in direction.

intelligent tutoring system (ITS) Systems that are designed to teach without a human tutor. The goal of these systems is to automate some pedagogical functions such as problem generation (new questions) and feedback generation (comments on the student's performance). Expensive to develop and implement, intelligent tutoring systems are believed to generate issues related to trust, mechanical interaction and emotions. It is not clear how effective they will be, compared to human teachers.

sensory modalities The ability to sense and control parameters related to physical contact with the external environment is a fundamental sensory capability required of robots. A sensory modality is one aspect of an external stimulus, such as light, sound, taste, pressure or smell. Machines have to learn to distinguish and interpret specific modalities out of the profusion of stimuli in the environment.

HUMAN-ROBOT INTERACTION

the 30-second data

HRI examines how people's attitudes and behaviours towards robots are affected by the robots' physical, cognitive and interactive features. The aim of HRI research is to make robots not only more efficient when working with humans, but also more acceptable. By definition, HRI focuses on the communication between one or more humans and one or more robots, which ideally is realized through several sensory modalities. A humanoid robot can interact through speech and facial gestures, whereas a floor-cleaning robot might use a simpler set of sounds (like R2D2 from *Star Wars*) to communicate. A mobile robot's navigation policy becomes a difficulty for HRI when working in public spaces: how should the robot approach people? Should it keep a distance when interacting with a human? How can it show the human its intention to pass? Owing to such human-related and un-measurable aspects of interaction, HRI methodology often relies on empirical analysis and qualitative evaluation. However, researchers also focus on establishing measurable and quantifiable benchmarks. As of the late 2010s, experimental protocols are typically limited to short-term laboratory studies. Hence a grand challenge of HRI for the long term is to enable robots to have autonomy in public spaces.

RELATED TOPICS
See also
ROBOTS FOR PEOPLE WITH SPECIAL NEEDS
page 84

TAMAGOTCHI
page 88

THE THREE LAWS OF ROBOTICS
page 144

3-SECOND BYTE
Human-Robot Interaction (HRI) research aims at promoting the relationship between humans and robots through a strictly multi-disciplinary approach, involving engineering, AI, psychology, sociology, cognitive science and philosophy.

3-MINUTE DEEP LEARNING
Ethical concerns go beyond safety and privacy: how should one define the robot's responsibility during emergencies, for example, for assistive robots in elderly care? How can you ensure the morality of artificially intelligent robots? How ethical is it to program a robot with human-like qualities? What happens if people develop attachments towards their robotic companions, which can break or be taken away?

3-SECOND BIOGRAPHIES
MAJA J. MATARIĆ
1965–
US pioneer in assistive robotics who works on HRI technologies for stroke patients, children with autism spectrum disorders, and people suffering from Alzheimer's disease and dementia

TAKAYUKI KANDA
1975–
Japanese computer scientist; he focuses on how robots should approach humans in public spaces, such as elementary schools, museums and shopping malls

30-SECOND TEXT
Ayse Kucukylimaz

Interaction means that we accept that robots act upon us.

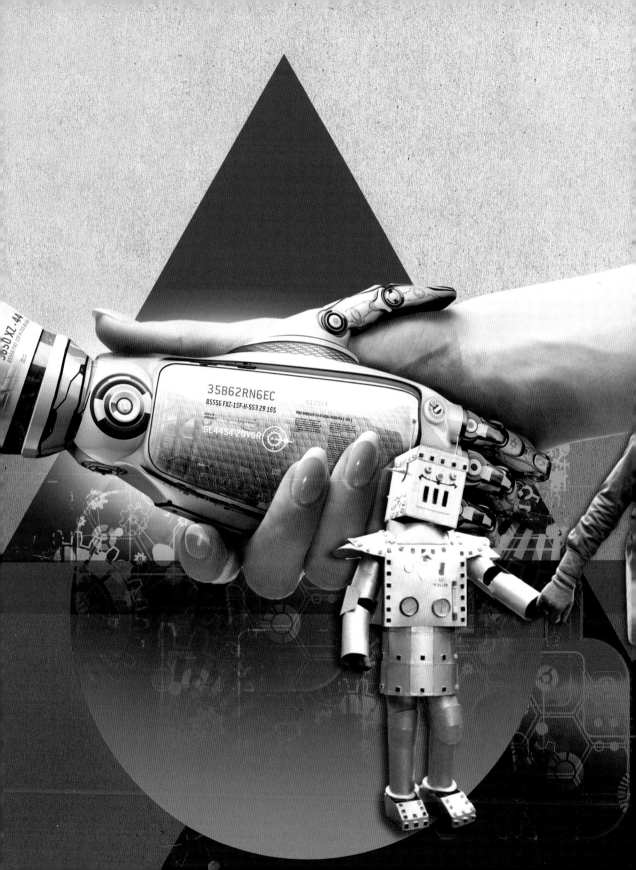

HOME ROBOTS & SMART HOMES
the 30-second data

3-SECOND BYTE
It's likely that future home robots will be integrated into smart-home settings – the whole environment will be designed as a network of AI systems to aid humans with household chores.

3-MINUTE DEEP LEARNING
'Domotics' means home automation. Early domotics technology focuses on connecting home devices over a sensor network for remote control and monitoring. The idea of the Internet of Things is to link these devices over a cloud network without the need for human interaction. More advanced domotics systems make use of artificial intelligence for the control of devices, while next-generation systems are expected to fully integrate robotic home assistants that can sense emotions.

Ever since the idea of robots emerged, humans have dreamt of domestic robots that vacuum, do the dishes and mow the grass. Despite this dream, as well as huge consumer interest and market potential, the use of sophisticated domestic robots remains limited. A pioneering and probably the best-known home robot is the Roomba series of autonomous robotic vacuum cleaners by iRobot. These robots can autonomously navigate your home to find dirty areas and sweep the floor using a pre-programmed routine. Some other smart-home solutions include robots for ironing and automatic window cleaning – each focusing on a single functionality. There are now two hotels in Japan that are almost entirely run by robots, which take on simple responsibilities such as carrying luggage, greeting humans and controlling TV functions, within a somewhat controlled environment. A major challenge for home robotics is that the robot must live with humans in a uniquely structured, cluttered and dynamically changing home environment. Smart-home technology already incorporates video monitoring, motion and fall detection, and environmental control. However, even within such a network, there are still many technical problems to be solved, such as 3D sensing, positioning, error recovery, safety, power requirements and multi-robot coordination.

RELATED TOPICS
See also
COLLECTIVE AI & CLOUD ROBOTICS
page 124

THE INTERNET OF THINGS
page 132

3-SECOND BIOGRAPHIES
JAMES (JIM) SUTHERLAND
fl. 1960s
American Westinghouse Corporation engineer who created the Electronic Computing Home Operator (ECHO IV) in 1966, considered to be the first home automation system

CYNTHIA LYNN BREAZEAL
1967–
American founder and Chief Scientist of Jibo, Inc., which produces Jibo, a smart home robot that functions as an interactive personal assistant

30-SECOND TEXT
Ayse Kucukylimaz

Tomorrow we will live inside a sophisticated robot.

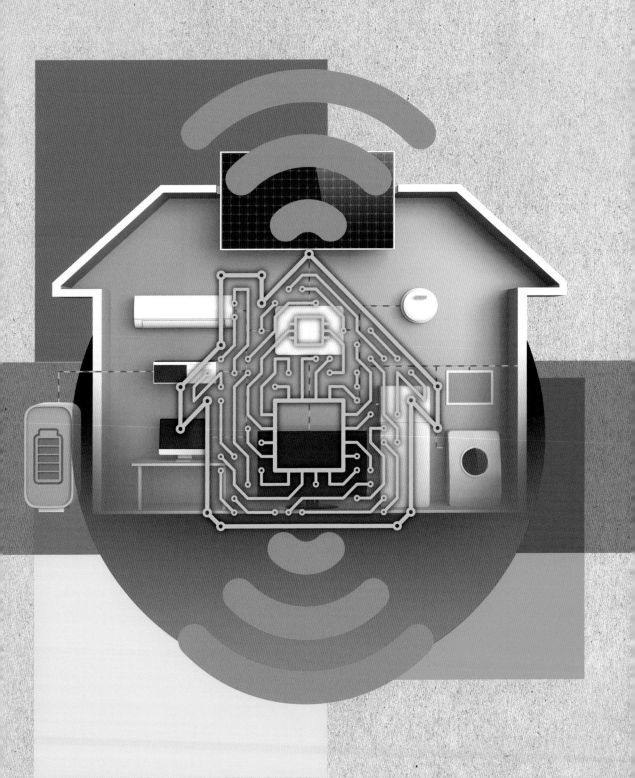

AUTONOMOUS CARS

the 30-second data

3-SECOND BYTE
Autonomous cars are equipped with multiple sensors and computers enabling them to navigate on public roads without the direct control of a human driver.

3-MINUTE DEEP LEARNING
The benefits of autonomous cars extend far beyond the convenience of individual owners. Vehicles capable of communicating with one another and the surrounding infrastructure could revolutionize the transport system. If vehicles on a motorway could share information about their destinations, they could operate as one large vehicle train, with each car driving faster and more safely than would otherwise be possible, and the reduced gaps between vehicles would vastly increase road capacity.

Traditionally, people expected autonomy in their vehicles. Horses came equipped with hazard recognition route following, and cruise control. As cars became popular, people tried to replicate these functions. The earliest systems were either controlled remotely or simply followed defined routes using buried magnetic guide wires. The earliest general-purpose autonomous vehicles were seen in the 1980s. In 2004, DARPA, the US Defense Research Agency, challenged teams to design an autonomous vehicle that could navigate a 240-km (150-mile) traffic-free route. Every entrant failed to complete the course; the best managed just 11.9 km (7.4 miles), but much was learned from the attempt. The following year, five teams successfully completed the course, and by 2007, autonomous vehicles were navigating complex urban environments with street signs and traffic. This success quickly led to the commercial adoption of the technology, and nowadays, several car manufacturers offer vehicles with varying degrees of autonomy. These cars use multiple sensors, typically Light Imaging, Detection and Ranging (lidar), visual cameras, ultrasound and inertial measurement sensors to create a dynamic model of everything around them. Eventually, autonomous cars may be capable of driving with no human supervision at all.

RELATED TOPICS
See also
GREY WALTER'S TURTLES
page 26

ROBOTS FOR PEOPLE WITH SPECIAL NEEDS
page 84

ELON MUSK
page 110

3-SECOND BIOGRAPHIES
NORMAN BEL GEDDES
1893–1958
American author of *Magic Motorways*, an early book that theorized about autonomous vehicles

FRANCIS P. HOUDINA
fl. 1920s
In 1925, Houdina demonstrated an early remote control car, named the 'American Wonder', in New York, USA

30-SECOND TEXT
David Rickmann

What will come first, the fully autonomous car or the end of physical travelling?

ROBOTS FOR PEOPLE WITH SPECIAL NEEDS

the 30-second data

3-SECOND BYTE
Robotic technology helps people with disabilities – it offers physical and social assistance and serves as a mental aid by facilitating learning and communication.

3-MINUTE DEEP LEARNING
The latest robotic developments could advance rehabilitation, but it gives rise to other concerns. People may become emotionally dependent on the machine and experience social isolation or the elderly may feel stigmatized; using a robot could draw attention to their disabilities. Prolonged use of robotic devices might turn patients into 'cyborgs' who consider the robot as part of their own bodies. Care decisions should therefore be based on the trade-off between care benefits and ethical concerns.

Certain people, such as the elderly and children with developmental disorders, need assistance with daily activities such as communication and eating. Robotic applications have been developed to support independent living. These new-age robots help share the daily burden of caregivers by taking on routine tasks such as reminding the elderly to take medicines on time. Robotic wheelchairs and walkers provide assistance with motor movements to patients recovering from brain damage. Similarly, children with autism spectrum disorders (ASD), who have difficulties with social interaction, have benefited from the introduction of anthropomorphic robots. Such robots are packed with simplified human-like features and are helpful in creating simple social situations for autistic children to explore social communication and learn socially acceptable behaviours. Furthermore, robots can also help reinforce classroom learning for children with special needs by offering them multiple opportunities for practising their lessons. Robotic technology should always be culturally adjusted. For instance, a culturally sensitive robot would not try to cheer up a recently widowed Indian woman as she is expected to mourn for at least one year. A robotic design that adheres to cultural norms is likely to be acceptable to a variety of users.

RELATED TOPICS
See also
UNIMATE, THE FIRST INDUSTRIAL ROBOT
page 56

HUMAN-ROBOT INTERACTION
page 78

HOME ROBOTS & SMART HOMES
page 80

3-SECOND BIOGRAPHIES
JOE JONES
1953–
American robotics visionary who created the Tertill, an autonomous robot that weeds the garden; he also helped to create the robotic vacuum cleaner, Roomba

DIWAKAR VAISH
1992–
Indian-born robotics researcher, creator of India's first humanoid robot, the 3D-printed Manav

30-SECOND TEXT
Neha Khetrapal

As robots become more autonomous, some of us will become totally dependent on them.

SMART ASSISTANTS & 'LITERATE' MACHINES

the 30-second data

3-SECOND BYTE
AI-powered digital assistants, also called smart or virtual assistants, are interactive software agents that can answer human daily requests, such as ordering a service vocally, providing information or entertainment.

3-MINUTE DEEP LEARNING
A perfect assistant should in theory be neutral and give you the best advice or point to the best possible service. But smart assistants are developed by multinationals that often have a financial interest in tailoring the information they retrieve for a specific user. Referring customers to commercial partners or apps is a way to generate revenue. Simultaneously, these software agents are collecting data on all aspects of our lives. Will we trust them?

The secretary might be a job of the past, but the idea behind smart assistants is that they will replace human assistants with an ever-available, all-listening artificial intelligence. Through voice recognition and natural-language processing, the assistant will be able to answer your questions and requests like an electronic butler, either embedded in your phone or embodied in a robot. Companies producing smart-assistant tools have been booming since 2015, with competing (often female) names such as Siri (Apple), Alexa (Amazon) or Cortana (Microsoft). Most digital assistants work with speech recognition, but voice might not be the main way we relate to smart AI in the future. As with mobile phones, text might come to predominate. Some companies are using deep learning techniques to focus on machine reading 'intuitive' comprehension. This is the ability to engineer systems that not only read entire documents but also answer increasingly complex questions about them. In a business context, the 'literate' assistant should be able to answer a question in a security-compliant manner through having a deep understanding of the contents of the organization's documents and e-mails, instead of simply retrieving a document by keyword matching. While extremely useful, this raises issues of privacy and confidentiality.

RELATED TOPICS
See also
METROPOLIS & THE GYNOID MARIA
page 20

EMBODIED AI & COGNITION
page 46

CHAT BOTS
page 92

3-SECOND BIOGRAPHIES
VANNEVAR BUSH
1890–1974
American computer scientist remembered for his 1945 article 'As We May Think', in which he foresaw what he called the Memex, a mechanized and archived form of intelligence

BARBARA GROSZ
1948–
American pioneer of natural-language processing and multi-agent systems, she works towards making human–computer interactions as fluent as possible

30-SECOND TEXT
Luis de Miranda

Our human assistants know a lot about us. Our smart assistants will know more about us than us.

October 1996
Consumer tests are done on high-school girls in Shibuya district of Tokyo with 200 prototype Tamagotchi units

23 November 1996
Tamagotchi is released in Japan

May 1997
Tamagotchi is released worldwide

August 1997
Tamagotchi Angel is released in Japan; it focuses on the afterlife of a Tamagotchi pet and includes motion and sound sensors for improved interactivity

1998
Tamagotchi Angel is released worldwide

2004
Tamagotchi Connection Version 1 is released worldwide; it includes a Tamacom infrared feature to connect two Tamagotchi pets

2013
Tamagotchi L.i.f.e. (Love Is Fun Everywhere) Android and iOS applications are launched by Sync Beatz Entertainment Inc.

November 2017
Tamagotchi Mini is relaunched as part of Tamagotchi's 20th anniversary celebrations

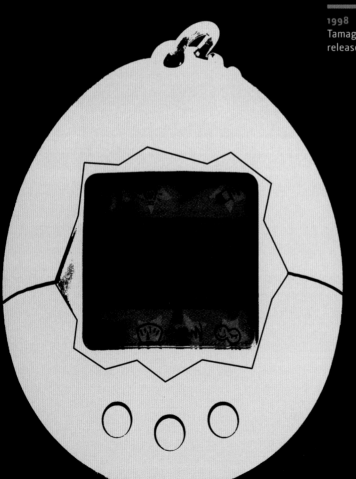

TAMAGOTCHI

Tamagotchi is a digital pet, living in an egg-shaped keychain with three buttons. The device allows a player to own an alien (Tamagotchi), ready to hatch from its egg. The aim of the game is to raise the creature through several stages of growth towards adulthood. Tamagotchi was conceptualized as an attention-seeking virtual pet that could help children to understand how to care for a real creature. Using buttons, the user can care for, feed and clean up after their Tamagotchi. The interface has a happiness meter and tools to monitor the creature's weight and health. The Tamagotchi can get hungry, defecate after eating, lose or gain weight, ask to play games, or fall ill. If the Tamagotchi is not attended to, it will 'die' of negligence.

Designed by Aki Maita and Akihiro Yokoi, Tamagotchi was released to the Japanese market by Bandai in 1996, prior to a worldwide release in 1997. The game achieved huge market success despite limited interactivity. In order to attract more users, Bandai's marketing strategy involved releasing new Tamagotchi versions almost every year with add-on features. In the 2010s, Tamagotchi applications for smartphones were launched. However, these attracted only limited attention from consumers, and did not achieve the success of the original version. In 2017, a 20-year anniversary version, Tamagotchi Mini, was released, striking a chord with those who cherished Tamagotchi when they were children back in the 1990s. According to Bandai, a total of more than 82 million Tamagotchi units have been sold worldwide, not including application downloads, and the virtual pet has been embraced by users of all ages.

The original Tamagotchi model used to die within half a day if left unattended. The very fragile and needy nature of the pet distracted child owners and caused emotional distress, and many schools banned the use of the toys. Newer versions attempted to address these issues. Yet being overly attached to Tamagotchis became identified as a condition called the 'Tamagotchi effect'. Discussion revolved around how humans can develop a connection with a non-living artefact when that object appears to give attention to them. Such attention allows humans to attribute 'intelligence' and 'emotions' to it. As in Tamagotchis, this effect is not related to whether the object can display intelligent or emotional behaviour, but rather to the fact that it can evoke emotions, which lead to attachment.

Ayse Kucukyilmaz

HUMANOID ROBOTS

the 30-second data

The earliest humanoids were built as early as the thirteenth century by Al-Jazari. In the fifteenth century, Italian scientist Leonardo da Vinci also designed a humanoid automaton. In Japan, mechanical dolls called Karakuri puppets were manufactured in the eighteenth century. However, these cannot be considered robots, because they lack the sensory system that is crucial for any mechanical system to be considered a robot. The first humanoid robot was called WABOT, and was developed by Ichiro Kato at Waseda University, Japan in 1970–73. One of the best-known humanoids, Honda's Asimo, was the product of a project that started in 1986. Honda robots were a leap forward in humanoid research due to their sturdiness, lightness and strength. Hiroshi Ishiguro from Osaka University is famous for his Geminoid series, which are very realistic androids. These robots are capable of human-like movements and gestures, even though their actions are controlled by remote human operators. When talking about humanoids, we immediately think of two-legged robots, and it is true that research on humanoid robotics is mostly focused on legged locomotion. However, robots consisting of only human-like upper bodies are also considered humanoids as their structure mimics the human form.

RELATED TOPICS
See also
KAREL ČAPEC & THE FIRST 'ROBOTS'
page 18

METROPOLIS & THE GYNOID MARIA
page 20

ISAAC ASIMOV
page 22

3-SECOND BIOGRAPHIES
ISMAIL AL-JAZARI
1136–1206
Arab scholar, inventor, and mathematician known for creating several mechanical inventions including automated humanoids, such as mechanical servants and a musical robot band

ICHIRO KATO
1925–94
Japanese professor who started the WABOT project in 1967 at Waseda University and built the world's first full-scale humanoid robot, WABOT-1

30-SECOND TEXT
Ayse Kucukyilmaz

A future in which we will have to prove daily that we are not a robot, until we can't.

CHAT BOTS

the 30-second data

The first famous chat bot was Weizenbaum's ELIZA (1964). ELIZA scans the user's input for particular words, and answers with pre-fabricated replies. If the user enters a sentence containing the word 'mother', ELIZA might reply with: 'Tell me more about your family.' A modern version of ELIZA is the chat bot ALICE. It contains around 41,000 patterns with suitable responses, and is available for a chat on the internet. Chat bots have been used experimentally for creative writing. William Chamberlain's program Racter (1983) produced stories and poems that sounded surprisingly human-like, and was credited as the author of a book of experimental poetry. Today, chat-bot research focuses on usability and commercial applications. Facebook and Twitter offer widely used platforms, where chat bots answer users' questions about particular products or services, and recommend products to buy. Chat bots have created a brand-new segment of online services. For the user, they provide an interface that is cleaner and easier to use than a website. But for the industry, chat bots strengthen the trend away from the branded websites of companies, and towards presenting every service as just an extension of Facebook or Twitter. In this way, they increase the dominance of social media as an information source.

3-SECOND BYTE
Using spoken language to access a computer has long been a feature of science fiction – now it has become reality in the form of chat interfaces.

3-MINUTE DEEP LEARNING
Researchers have been trying to create convincing chat bots ever since Alan Turing proposed a human-like chat as a test for machine intelligence. But as ELIZA and ALICE show, fooling humans into attributing intelligence to a program is surprisingly easy. Many other tasks, such as driving an autonomous car or understanding written language, require far more complex programs and more resources. So perhaps human-like chat is not the holy grail of AI after all.

RELATED TOPICS
See also
HAL 9000 IN *2001: A SPACE ODYSSEY*
page 24

THE TURING TEST
page 34

LANGUAGE RECOGNITION & TRANSLATION
page 44

3-SECOND BIOGRAPHIES
TERRY ALLEN WINOGRAD
1946–
American professor of computer science who created one of the first programs to understand natural language, SHRDLU

SHERRY TURKLE
1948–
American professor at MIT who has warned that increasing communication with machines might reduce humans' ability for social interaction

30-SECOND TEXT
Andreas Matthias

Chat bots offer a 'human' service online, but everything you say may be used for or against you.

AI IN EDUCATION

the 30-second data

Artificial Intelligence systems can be helpful to educators by taking up some of the more routine tasks such as marking, leaving teachers more time to teach. While multiple-choice tests are simple to grade, assessing an essay requires an understanding of both the subject and the nuance of a written argument. In 2012, the William and Floran Hewlett Foundation offered a US $60,000 prize for an automatic essay-grading system. Teachers marked a set of essays that were also marked by AI systems. The winning AI team gave 81 per cent of papers the same grade as was given by the teachers. By 2016 that figure had been raised to 94.5 per cent. AI can also assist educators with their administrative tasks. In 2016, Professor Ashok Goel at the Georgia Institute of Technology designed 'Jill Watson' to serve as a teaching assistant for his online course, *Knowledge-Based Artificial Intelligence*. Jill can respond to routine questions on such things as appropriate file formats and office-hours timetables, while leaving more detailed or creative questions to Goel and his human team. Ideally, teachers would give students one-to-one instruction as required, but large class sizes make this impossible. This leads to a significant achievement gap between students who can afford personal tutoring and those who cannot.

RELATED TOPICS
See also
LANGUAGE RECOGNITION
& TRANSLATION
page 44

INTELLIGENCE
AMPLIFICATION
page 106

EXTENDED MINDS
page 146

3-SECOND BIOGRAPHIES
DONALD L. BITZER
1934–
American electrical engineer; in 1960 he designed PLATO (Programmed Logic for Automatic Teaching Operations), a computer-based education system

HIROSHI KOBAYASHI
1966–
In 2009 Professor Kobayashi headed the team at the Tokyo University of Science that developed and deployed Saya, the first humanoid robot teacher

30-SECOND TEXT
Lisa McNulty

3-SECOND BYTE
AI can improve learning and teaching by automating assessment, assisting teachers with routine tasks, and providing 'intelligent tutoring systems' that meet students' individual needs.

3-MINUTE DEEP LEARNING
While an ITS can help bring individual tutoring to a large student body, there are limitations to the approach. A skilled teacher is able to identify when their students are confused or frustrated and adapt their methods to address these emotional states. The next evolution of the ITS is the Affective Tutoring System (ATS), an AI system that can read facial expressions, body gestures and speech in order to identify and adapt to the student's state of mind.

Will artificial tutors respond to essential individual needs?

CAN MACHINES HAVE COMMON SENSE?

the 30-second data

3-SECOND BYTE
Machines with common sense could understand everyday language, make better decisions and avoid many mistakes. But how can we build them?

3-MINUTE DEEP LEARNING
The so-called 'frame problem' of AI is an extension of the common-sense problem. If I paint a cupboard in my kitchen, which properties of the world are going to change? Obviously, the colour of the cupboard. But also the smell of my kitchen, the weight of the paint pot, the state of the brush, and so on. How could a machine know all the consequences of an action, even such a simple one?

What is common sense? It is

a kind of 'background knowledge' that lies behind our understanding of the world. It is the stuff that we don't need to explain, because it's obvious to everyone who shares our 'common' sense. Machines today entirely lack common sense. Calendar apps will happily book a dinner in a steakhouse for a vegetarian. Also, different cultures, and even a farmer compared to an office worker, will have different common sense. One needs to grow up in a particular culture to acquire a shared common sense. Language understanding often depends on common sense, too: If I say 'Mary saw the dog in the shop window and wanted it,' you need common sense to understand that Mary wanted the dog, not the window. How could a machine possibly acquire such common sense? One way would be to provide the machine with common sense in the form of facts and rules (as in a computer program). The CYC research project (since 1984) has attempted to do just that. A knowledge base released in 2012 contained 239,000 concepts and 2,093,000 facts describing human background knowledge. Because of the immense number of facts that need to be input into the database, after over 30 years, the CYC project is still far from completion.

RELATED TOPICS
See also
RODNEY BROOKS
page 64

HUMAN-ROBOT INTERACTION
page 78

SMART ASSISTANTS & 'LITERATE' MACHINES
page 86

HUBERT DREYFUS
page 148

3-SECOND BIOGRAPHIES
MAURICE MERLEAU-PONTY
1908–61
French philosopher who argued that the body plays a central role in perception and knowledge

DOUGLAS LENAT
1950–
American founder and principal researcher of the CYC project, and CEO of Cycorp, the company behind CYC research

30-SECOND TEXT
Andreas Matthias

What if common sense is a form of common decency? And how can machines acquire it?

CYBORGS & AUGMENTED HUMANS

CYBORGS & AUGMENTED HUMANS
GLOSSARY

adjuvant Substance used in combination with an antigen treatment to improve its desired effect, for example, to enhance the speed and duration of an immune response. AI machine-learning algorithms have been developed that use statistical tools to predict outcomes, for example, to evaluate the prognosis in breast-cancer cases and recommend the best adjuvants.

augmented reality Brings elements of the virtual world into our perception of reality, adding information to what we see or experience and creating mixed reality. The aim is to bring computer-generated objects or information into the real world, which only the user can see. A doctor might use augmented-reality glasses to access your medical records while talking to you.

bioethicist A specialist who applies ethical questions to the field of life manipulation or preservation. This can include pharmaceutical drugs, wildlife conservation, or human health and treatment, such as abortion or euthanasia. In medicine, the autonomy of patients is a bioethical issue: do they have any freedom to choose their own treatment?

bio-hybrid Bio-hybrid robots (or biobots) are made by combining robotics with tissue engineering, adding muscle or cells to the machine structure. These devices can be stimulated electronically or with light, for example, to make the cells contract to bend their skeletons, causing the robot to swim or crawl. Bio-hybrid robots are often inspired by animals, such as jellyfish.

biomechatronics Machines that are made of mechanical parts, electronic parts and computer elements. Mechatronic engineers work on solving the integration of these various approaches. Today, a fourth component is added to the mix, the biological, and a new interdisciplinary study has emerged, which is probably the future of robotics: biomechatronics. This is the endeavour to build bio-hybrid robots that are human-like, and in which nerve cells or even neurons are connected to electronic and mechanical parts.

biomimetic Refers to the process of mimicking living plants or animals to emulate the way in which they solve problems or tackle tasks. Biomimetic robots are expected to exhibit greater competence than regular robots in environments where a certain element of improvisation is needed. The goal is to create machines that are semi-autonomous and flexible.

biotechnology Technology based on a living organism, which harnesses cellular and biomolecular processes to make a useful product, such as a vaccine, an antibiotic or a synthetic organism. Many forms of contemporary biotechnology rely on DNA technology. Biotechnology inventions can raise ethical concerns, for example, about life manipulation, food quality or privacy.

CRISPR technique Pronounced 'crisper', an abbreviation for Clustered Regularly Interspaced Short Palindromic Repeats, a technology that since 2013 has allowed easy and precise genome editing. Scientists can rapidly recreate cells and biological models, for example, to produce mutations. Some people, especially in China, are having their genes edited using CRISPR, although this raises bioethical concerns.

dermal Relating to the skin. A key focus for the field of biorobotics is to work on the reproduction of skin-like envelopes, which would allow for a kind of sensing analogous to human or animal sensing. Dermal sensors, which would comprise dozens of sensing zones, are intended to mimic the role of skin receptors.

nanotechnology The study of extremely small things (1 to 100 nanometres) with the tools of chemistry, biology, physics or engineering in order to control individual atoms or small groups of atoms. Materials at this scale often have distinctive properties; the aim is to manipulate them into new forms and functions, for example, for medical purposes.

psycho-pharmacological To do with the study of the effects of drugs and other substances on the mind, including thoughts, sensations and certain behaviours.

synthetic biology Direct access to the genetic code of animals, plants or humans allows several types of DNA manipulation, through genetic engineering. Synthetic biology is the idea that we will transform nature via a re-coding of DNA structures, which will then reproduce naturally. One application is to make artificial food, such as synthetic meat.

HUMAN ENHANCEMENT

the 30-second data

How and why should we improve ourselves? This question might not be new, but the widespread use of biotechnologies in medical sciences gives it a sense of urgency. From computer brain implants to nanotechnology in cosmetic surgery, the debate is vigorous between those who think we should modify humans to 'perfect' them and those who believe it is unethical and dangerous. Human lives have always been exposed to the unbidden and to the latter it feels like sacrilege to control everything by manipulating our genes, brains, bones and muscles. Yet the World Health Organization defines health as 'a state of complete physical, mental and social well-being and not merely the absence of disease or infirmity'. This means that non-medical problems, such as ageing or shyness, may come to be defined and treated as matters for technological or chemical improvement. Transhumanists believe that being a modified cyborg is only the next stage of our evolution, as we go beyond our human limitations. Yet bioethicists point to security and societal side effects, such as increased inequality (richer people having easier access to enhancements which might make them even richer) and the pathologies of competitive pressure. Despite the controversy, human enhancement is becoming a major economic market.

3-SECOND BYTE

Augmenting human performance, appearance or behaviour through genetic science, medicine and technology is one of the most fascinating promises of current technologies. Is it for better or worse?

3-MINUTE DEEP LEARNING

What makes a human being authentic? Aldous Huxley's *Brave New World* is the perfect metaphor for reflecting on the societal effects of large-scale human enhancement. In the novel, citizens are engineered through artificial wombs, selective breeding and psycho-pharmacological substances, in a society divided into castes. People get what they want because they never want what they can't think of.

RELATED TOPICS

See also
HUMANOID ROBOTS
page 90

TRANSHUMANISM & SINGULARITY
page 116

3-SECOND BIOGRAPHIES

HANS BERGER
1873–1941
German psychiatrist who invented the electroencephalograph (EEG), which records brain activity; the idea that the brain is mainly an electrical device is key to current work on thought-controlled robots

ALDOUS HUXLEY
1894–1963
The British author of *Brave New World* was a critic of our modern democratic life, in which human enhancement seems to create extreme conformity

30-SECOND TEXT
Luis de Miranda

Aldous Huxley felt that a systematic enhancement of our social species could eliminate individual differences.

BIONIC PROSTHETICS & EXOSKELETONS

the 30-second data

3-SECOND BYTE
Biomechatronic robotic systems are developed to replace or enhance human functionality, typically for mobility and manipulation. Prostheses are artificial limbs, while exoskeletons (orthoses) are used for human augmentation.

3-MINUTE DEEP LEARNING
Every year, the number of first-time stroke incidents worldwide is more than 15 million. Most stroke survivors require long rehabilitation therapies, which are physically demanding for them and for their therapists. Therefore, developing active exoskeletons for rehabilitation is an important research area. Another application for exoskeletons is medical care, where exoskeletons are worn by the nurses to help them lift and carry patients.

A prosthesis is a functional replacement device that is used in exchange for a missing body part. Integrating these devices into the body and making them transparently controllable by the human is a complex task. Prosthetic devices make use of data coming from sensors located on the outside or the inside of a user's body, to interpret the motion intentions of the user. Special techniques can be used to surgically reroute certain motor nerves to intact muscles. A popular type of sensor detects muscle activity, and is called a myoelectric sensor. Myoelectric sensors are widely used due to their robustness. Electrodes installed directly into the brain have been shown to be useful when carrying out complex tasks, such as grasping and two-handed control of two prosthetic arms. On the other hand, exoskeletons are wearable devices, which do not replace but augment the non-functioning limb. For example, a paralysed person can wear an exoskeleton to provide active or passive walking support. Several types of exoskeletons are manufactured, mainly for medical purposes. Unlike prosthetics, exoskeletons can be used for human performance augmentation. Sometimes dubbed powered suits, a huge application area for such exoskeletons is the military.

RELATED TOPICS
See also
ROBOTS FOR PEOPLE WITH SPECIAL NEEDS
page 84

HUMAN ENHANCEMENT
page 102

INTELLIGENCE AMPLIFICATION
page 106

3-SECOND BIOGRAPHIES
KEVIN WARWICK
1954–
British engineer, known as 'Captain Cyborg', who installed an electrode array on himself to remotely control a robot arm

HUGH HERR
1964–
American engineer, biophysicist and amputee, who designed a specialized prosthesis for his legs that allows him to perform rock climbing under neural commands

30-SECOND TEXT
Ayse Kucukyilmaz

Humans have always had a problem with their bodies and the temptation to modify them.

INTELLIGENCE AMPLIFICATION

the 30-second data

As computing technology

develops the capacity to imitate human cognition, two competing approaches have emerged. AI delegates cognitive processes solely to machines, while IA augments the brain to enhance its cognitive capacity. This might be by automatically translating languages, enhancing facial recognition or even turning on the lights with a thought. Just as we use technology to enhance strength, we can use it to enhance thinking, thus creating a collaboration between the brain and external devices or adjuvants. An early version of IA was developed in 1993 by American computer scientist Thad Starner (1970–), in the form of a wearable computer assistant called Lizzy. He used it so extensively that his PhD committee considered awarding his doctorate to the combined entity 'Thad Starner and Lizzy'. For many people, their computer or smartphone performs a similar function today. The idealized version of IA is a computer and brain, directly linked. This idea rose to prominence in the 1960s, when J. C. R. Licklider theorized about coupling a computer system with a human brain via a direct neural connection. He suggested that mundane cognitive tasks would be offloaded to the computer, for example calculation or statistics, thus liberating brain time for more creative forms of cogitation and awareness.

If our brain was connected to the Internet, could it process information?

GENETIC ENGINEERING & BIOROBOTICS

the 30-second data

the 30-second data

3-SECOND BYTE
Biorobotics is the attempt to build robotic devices with partly biological parts, produced via genetic engineering to simulate or merge with living organisms.

3-MINUTE DEEP LEARNING
Biorobotics is not only about making partly biological robots; it is also a therapeutic and medical field. Bioengineered parts could be created to cure diseases or injuries using artificial sensing skins, limb prostheses, dermally implanted sensors, cellular surgery or implantable devices. Wearable electronic tattoos that monitor bodily electric signals could cure sleep disorders or the heart activity of premature babies. The twenty-first century is likely to be the century of biorobotics.

We are familiar with the idea –

at least since the classic movie *Blade Runner* (1982) – that one day robots might be fully alive. The science of genetic engineering is already able to create organisms via artificial means using biotechnologies. The first genetically modified (GM) organisms were bacteria (1973); the first GM animal was a mouse (1974). In 2010, the first synthetic DNA genome was inserted into a cell that could replicate and produce proteins. Since 2013, the CRISPR technique has made it possible to easily manipulate the genome of many organisms. It seems that animal-like robots will be the missing link between the realms of biology and engineering. Scientists are using biomimetic approaches – that imitate nature – to create mechanical salamanders, dogs or spiders that move like real animals. From imitation to reality, there is just one step. In 2012, researchers from Newcastle University in the UK engineered protein cells from a Chinese hamster ovary to respond to visible light. It can convert an incoming optical signal to a chemical signal, which could then be converted into an electrical signal. Their goal is to create a swimming bio-hybrid robot: photosensitive engineered cells will be the 'eyes'. The basic idea of biorobotics is that life is a combination of electrical exchanges, so there is continuity between animals and machines.

RELATED TOPICS
See also
NANOBOTS
page 130

THE MERGING OF ROBOTS,
AI & HUMANS?
page 150

3-SECOND BIOGRAPHIES
JAMES WATSON & FRANCIS CRICK
1928– & 1916–2004
American and British biophysicists famous for discovering the double-helix structure of DNA in 1953; knowing how DNA replicates was the first step towards its modification

JOHN CRAIG VENTER
1946–
American biotechnologist and controversial businessman, he was instrumental in the sequencing of a human genome in 2007. He now conducts research in synthetic biology

30-SECOND TEXT
Luis de Miranda

If everything is about electrical exchanges, even in our bodies of flesh, then we will easily merge with technology.

28 June 1971
Born in Pretoria,
Transvaal, South Africa

1992–5
Attends the University
of Pennsylvania

1995
With his brother, founds
Zip2, a web software
company

1999
Co-founds X.com, an
online payment company
that later merges with
Confinity and becomes
PayPal

2002
Becomes a US citizen and
founds SpaceX

2008
Becomes CEO and
product architect of
Tesla Motors

2013
Develops concept
for a high-speed
transportation system
called Hyperloop

2015
Creation of OpenAI, a
non-profit AI research
institute

2016
Founds Neuralink,
a neurotechnology
start-up, aiming to
connect the brain
with AI technology

2016
Founds the Boring
Company, with the
goal of creating more
underground traffic
tunnels

2018
Musk's frequent polemical
statements and the
financial uncertainties of
his companies lead to
criticism in the media

ELON MUSK

Elon Musk's life looks as if a crazy biographer mixed up several different lives, pouring all their events into a single timeline. As a child, he was an enthusiastic book reader, who was sometimes bullied and once thrown down a flight of stairs and beaten until he lost consciousness. At age 12, he taught himself computer programming and sold his first computer game to a magazine. After studying in Canada for two years, he moved to the United States, where he graduated in both Physics and Economics.

His first company was a web software start-up, Zip2. When Musk sold it, he received US $22 million from his shares. Using a part of this money, he co-founded X.com, which later became the online payments company PayPal. From then on, he built a series of companies in different areas, all motivated by the basic question: 'what do I think is going to most affect the future of humanity?' (CalTech commencement address, 2012).

Musk decided to tackle the dependence of transportation technology on fossil fuels. From 2003, he worked for Tesla, just as the company introduced electric cars to mainstream consumers. To meet Tesla's need for powerful batteries, he built a battery 'Gigafactory', and, as a by-product, a line of household batteries that can power a home. To recharge batteries in an environmentally friendly way, Musk provided the concept and initial capital for SolarCity, now the second-largest US manufacturer of solar-power systems.

Believing that being limited to a single planet is a danger to humanity, Musk established SpaceX. The company has built the first reusable orbital rockets, transported goods to the International Space Station, constructed the most powerful rocket engines to date, and is planning a voyage to Mars, with the future aim to colonize the planet.

Still unhappy about transportation options, Musk set up Hyperloop, a company that hopes to build a high-speed train inside a low-pressure tube, and the Boring Company, a start-up aiming to construct subterranean transportation tunnels.

In 2015, as AI and its risks rose to prominence in the media, Musk warned of the potential of AI to destroy human civilization. He founded the OpenAI institute, which intends to undertake AI research in a way that is safe and beneficial to humanity. He also set up Neuralink, which researches methods to connect the human brain with computers, bypassing inefficient traditional input–output devices, such as monitors and keyboards.

Andreas Matthias

VIRTUAL REALITY

the 30-second data

3-SECOND BYTE
Virtual reality is a computer-generated 3D simulation or recreation of a real-world environment, usually used in gaming, education, exploration or marketing experiences.

3-MINUTE DEEP LEARNING
Imagine walking down a street with extra information about your surroundings popping up in front of your eyes. If you combine the perception of the real world with superimposed elements of VR, you get 'augmented reality', which is different from virtual reality because it does not replace your familiar world entirely. Real objects are equipped with sensors and transmitters that communicate remotely with a headset device. One day, they could link directly to your brain via electronic waves and chips.

We are familiar with being so absorbed in a movie that it feels we have been teleported into the story, alongside the protagonists. This is because our brain tends to identify what it sees and hears with reality. Back in 1838, Charles Wheatstone's stereoscopic photos showed that the brain unifies the slightly different two-dimensional images from each eye into a single object of three dimensions. In the 1950s, Morton Heilig developed the Sensorama, an arcade-style cabinet stimulating all the senses. Today, virtual reality is associated with head-mounted displays (HMD), giving a stereoscopic 3D effect of a computer-generated milieu. You are immersed in a parallel world, and the movements of your head or body affect what you see, whether it's a magic castle or a virtual shop. AI is now starting to be associated with VR to create a sense of improvisation that feels more personal to the user. Brain–computer interface technologies for VR and robotics are progressing quickly. Some computer engineers want to connect the headset to our brain so that the simulated reality responds to our emotions, intentions or even thoughts. This technology could be applied in health situations (re-embodiment after injuries), work environments (controlling a robot with a VR display) or in online shops ('feeling' a product on you before you buy it).

RELATED TOPICS
See also
EMBODIED AI & COGNITION
page 46

THE INTERNET OF THINGS
page 132

EXTENDED MINDS
page 146

3-SECOND BIOGRAPHIES
CHARLES WHEATSTONE
1802–75
British inventor, considered the father of 3D and VR for his use of stereoscopy to understand how the brain perceives the world

MORTON HEILIG
1926–97
American pioneer in VR, he patented the Sensorama in 1962, an immersive 3D-view experience, which he hoped would be the cinema of the future

30-SECOND TEXT
Luis de Miranda

Given the time we already spend in front of 2D screens, it is likely that virtual reality will be very addictive.

MIND UPLOADING

the 30-second data

3-SECOND BYTE
What if we could upload our minds to a computer and live forever inside its memory – and perhaps, one day, find a second life in another body?

3-MINUTE DEEP LEARNING
Brain uploads bring with them a whole array of legal problems. For example, when we simulate the whole brain of a mouse in a computer, does this simulated brain also feel pain? Philosophers such as Tyler Bancroft have argued that such simulations should be granted animal rights and be protected from suffering. And if a dying man uploads his consciousness into a computer, should this simulation of his mind also inherit his property and social position?

Every lesson learned is a transfer of mental content from one person to another. Will it be possible to transfer our consciousness in the same way? Everything we know is stored in the 90 billion neurons of our brain. Research projects such as the Blue Brain project aim to map these neurons, and to contribute to a complete map of the human brain. For the time being, we have only one such map, depicting the 300 neurons of the small worm *Caenorhabditis elegans*. Inventor Raymond Kurzweil believes complete brain uploads to a computer will be possible by 2045, making human minds immortal. The problem is that nobody knows how much of our consciousness is stored in the neural structure of our brains. It might be that there is some particular 'mind substance' or 'soul' that is necessary for the mind to work, as Descartes and many philosophers have believed. It could be that the electrical and chemical activity of the brain must also be captured, along with the neural structure, which would be technically difficult. Scientists have found strong correlations between brain structure and behaviour traits, such as anger or rule breaking. Such findings strengthen the argument that one day, by decoding brain structure, we might be able to transfer at least some aspects of human consciousness to a machine.

RELATED TOPICS
See also
EMBODIED AI & COGNITION
page 46

ELON MUSK
page 110

3-SECOND BIOGRAPHIES
RAYMOND KURZWEIL
1948–
American computer scientist and futurist who has predicted that within the twenty-first century, computers will be able to operate in the same way as the human brain

CHRISTOF KOCH
1956–
American neuroscientist, known for his research into how consciousness can come about in information systems

RANDAL A. KOENE
1971–
Dutch neuroscientist and founder of a project that aims to advance mind uploading

30-SECOND TEXT
Andreas Matthias

What if we could upload our childhood mind and reconnect with it once we are an adult?

TRANSHUMANISM & SINGULARITY

the 30-second data

Computing power has increased exponentially over the past decades. Today, AI systems can do things that were considered science fiction 20 years ago. Computers can play chess and Go better than humans, they can diagnose diseases and drive cars. If computers keep getting faster and more intelligent, will there come a time where humans will look like primitive barbarians in comparison? From then on, computers will be able to design new computers without the help of humans, further increasing the gap between man and machine. Science-fiction writer Vernor Vinge coined the term 'singularity' for the time at which computers become more intelligent than humans. Should the singularity come about, there seem to be only two options for humans: first, we could accept our fate and live a life controlled by super-intelligent (and hopefully benevolent) AI systems. Alternatively, we could try to change ourselves to match the capabilities of our computers. This is the approach of transhumanism. With the integration of new sensors and computational modules into our bodies, and by interfacing our brains with computers, we might be able to develop new abilities that will fundamentally change the human condition. We might become immortal by uploading our minds into computers, or acquire the ability to instantly understand every human.

RELATED TOPICS
See also
HUMAN ENHANCEMENT
page 102

MIND UPLOADING
page 114

3-SECOND BYTE
If computers become more and more intelligent, they are likely to reach a point where their intelligence will surpass ours – but what will happen after that?

3-MINUTE DEEP LEARNING
Since the Renaissance, humanism has sought to emphasize the aspects of humanity that are shared. It assumes that what unites us is stronger and more significant than our differences. Transhumanism, in contrast, advocates improving the human condition through the use of technology. But these technologies are not available to all, and thus transhumanism is in danger of creating new inequalities between those who can afford access to its promised land and those who cannot.

3-SECOND BIOGRAPHIES
FM-2030
1930–2000
Belgian-born transhumanist philosopher, who adopted this name to free himself from what he considered 'tribal' mentality

HANS PETER MORAVEC
1948–
Austrian-born Canadian computer scientist, technologist and writer about transhumanism

NICK BOSTROM
1973–
Swedish philosopher who warned of the dangers that AI might pose, and emphasized the need for better control of AI systems

30-SECOND TEXT
Andreas Matthias

Transhumanists believe we need to mutate into a new species, made of flesh and digital parts.

COLLECTIVE ROBOTICS & AI

actuator This is a device that acts on something else, usually stimulating a movement, such as opening a door or ejecting an object. In robotics, actuators are used to facilitate the movement of the robot parts, for example, rotation.

ARPANET Acronym for Advanced Research Projects Agency Network, a body of the US Defense Department. Developed in the late 1960s, this computer network was the forerunner of the internet. It connected the computers of Pentagon-funded institutions via telephone lines.

backpropagation Short for 'backward propagation of errors', this is an extremely efficient logarithm that is used in neural networks and machine learning: the formula assesses the level of mistakes made by a neural node, and constantly improves its efficiency.

biomaterial A substance that has been engineered to take a form that interacts with living organisms. Typical applications are joint replacements, heart valves, breast and dental implants.

crowdsource To outsource work or a problem to an unspecified group of people via the internet. It can be used for scientific and non-profit causes, for example to analyse large quantities of data or to validate information.

epigenetic The study of how genes can be influenced by the environment or lifestyle of the body they express. For example, human epidemiological studies appear to provide evidence that pre-natal and early post-natal environmental factors influence the risk of developing various chronic diseases and behavioural disorders in adulthood.

evo-devo Abbreviation of evolutionary developmental biology, a field of biological research that studies the interaction between developmental processes and evolutionary factors, and the variations generated by the former on the latter.

genome The collection of all DNA in chromosomes and cells of a given living organism. An average human has ten trillion cells in their body in which DNA is coiled. In 2003, the Human Genome Project identified the complete sequence of our genetic code.

hive mind In computing, the collective thoughts, ideas and creativity of a group of people, such as internet users. The hive-mind theory suggests that the sum of individuals' efforts is much greater than what each person would be able to achieve.

law of large numbers In statistics, the theory that the larger the sample population used in a test, the more accurate the prediction of the behaviour of that sample will be. It is also called the law of averages.

nanometre A unit of length in the metric system, equal to one billionth of a metre (0,000000001 m). The word is built on the Ancient Greek root nano, meaning 'dwarf'. This measure is often used to express dimensions on an atomic scale.

nanotechnology The study of extremely small things (1 to 100 nanometres) with the tools of chemistry, biology, physics or engineering in order to control individual atoms or small groups of atoms. Materials at this scale often have distinctive properties; the aim is to manipulate them into new forms and functions, for example, for medical purposes.

neuroevolution A sub-field within AI and machine learning, the aim is to emulate via computer programs an evolutionary process similar to the one that produced our brains over millions of years.

phenotype While the genotype is the complete heritable genetic identity of a person, the phenotype is a description of the actual physical characteristics of that person, such as eye colour and blood group.

photolithography Also called optical lithography, a method that allows the production of microprocessors and electronic circuits.

RFID tag RFID tagging uses small radio frequency to identify devices and track them, often using a chip with an antenna. It is used as an alternative to barcode technology, for example in supply-chain management.

stigmergy Refers to the idea of swarm intelligence in social insects: complex structures can be built by unintelligent individuals communicating with each other. Traces of chemicals (such as pheromones) left in the environment or modifications made by individuals are used as feedback. The colony uses the physical environment as a set of instructions for the next step, somewhat like an algorithm.

MULTI-ROBOT SYSTEMS & ROBOT SWARMS

the 30-second data

Multi-robot systems appeared

in the research literature in the 1980s, as an extension of single-robot systems. MRS are often cooperative systems, meaning that the team members work jointly to solve a problem. Each team member is assigned a role that it must carry out in coordination with the rest of the team. The robots are located in a particular formation, depending on the environment, and they may need to switch between different formations over time. To be able to cooperate, the robots act as mobile sensors, observing targets, mapping the surrounding area and sharing the information. Each robot's plans are synchronized with the others to ensure they cooperate effectively to achieve their goals. There are many applications of MRS. They can be used for information gathering and communications: the surveillance and/or exploration of large areas, the tracking of targets and re-configurable wireless communication networks. Robots can carry out physical tasks, such as cooperative manipulation, cooperative transportation, and search and rescue operations involving aerial and ground robots. Nanorobots are used for the distributed sensing of the human body. The applications of MRS and their interaction with human groups are expected to increase steadily in the upcoming years.

RELATED TOPICS
See also
COLLECTIVE AI & CLOUD ROBOTICS
page 124

EVOLUTIONARY AI & ROBOTICS
page 134

3-SECOND BYTE
Multi-Robot Systems (MRS) are robot collectives, composed of a variable number of robots operating cooperatively to achieve goals that would be difficult or impossible for a single robot.

3-MINUTE DEEP LEARNING
Robot swarms are MRS composed of a large number of robots. Each robot exhibits simple behaviours, but complex collective behaviours emerge from the interactions between swarm members. The interaction rules are often inspired by those used by animal collectives. Biologists have studied animal collectives to explain their complex behaviours. MRS designers study the animal interactions in order to create bio-inspired swarm robots.

3-SECOND BIOGRAPHIES
PIERRE-PAUL GRASSÉ
1895–1985
Influential French zoologist who introduced the concept of stigmergy, a method for coordination adopted by social insects, utilizing traces left in the environment

CRAIG REYNOLDS
1953–
American expert on computer graphics, famous for the creation (in 1986) of Boids, an artificial life simulator. Its flocking rules for bird-like animated computer-graphic objects inspired formation control methods for robot swarms

30-SECOND TEXT
Pedro U. Lima

How to achieve a maximal effect with disposable individuals: a lesson inspired by nature.

COLLECTIVE AI
& CLOUD ROBOTICS

the 30-second data

Ants work together to build

complex structures. No individual is in charge, nor does any ant understand the whole system; organization emerges from an invisible collective memory. Collective AI and cloud robotics follow this analogy: instead of building highly autonomous units (computers, software or robots), they take advantage of the law of large numbers and online big data. Analytic power is crowdsourced – 'outsourced to the crowd' – which means that the source of knowledge is not in the individual but in groups of humans and data. First, we record and share thousands of performances, say, a collection of past medical diagnoses. These are mined for logical patterns (for example, symptoms associated with a particular disease). The results can help in the diagnosis of a new individual case. Similarly, in cloud robotics, rather than implementing AI technology into one fixed program, robots have simple algorithms that draw from online information. The 'cloud' (internet-hosted services and datasets) has constantly evolving information: libraries, images, maps, models, habits, recipes and other datasets. Thus cloud intelligence relies on a grid of connected computers, like an artificial collective memory. The convergence of advances in global connectivity, storage and processing power strengthens the power of collective AI.

RELATED TOPICS
See also
SMART ASSISTANTS &
'LITERATE' MACHINES
page 86

MULTI-ROBOT SYSTEMS
& ROBOT SWARMS
page 122

EXTENDED MINDS
page 146

3-SECOND BIOGRAPHIES
ÉMILE DURKHEIM
1858–1917
French sociologist; proposed the notion of collective consciousness to explain why groups outperform individuals

HERBERT ALEXANDER SIMON
1916–2001
American sociologist, economist and computer scientist known for his work on the theory of corporate decision making. He was a pioneer in creating AI through computer technology

30-SECOND TEXT
Luis de Miranda

Once everything is interconnected, it becomes a collective being, with no centre and no limits.

3-SECOND BYTE
Collective AI simulates intelligence using big data accumulated online, tapping into the patterns that are generated by the experience of many humans or hundreds of machines.

3-MINUTE DEEP LEARNING
Can robots learn from robots? Take all the skills and specialities that each of them has, push all the information online and share it. Having this information on the 'robot cloud' means that new machines will be able to tap into all the gathered learning. Over time, as a given company's robots are trained and train others, they will become better at identifying the most important information available on the cloud servers. Will they still need us?

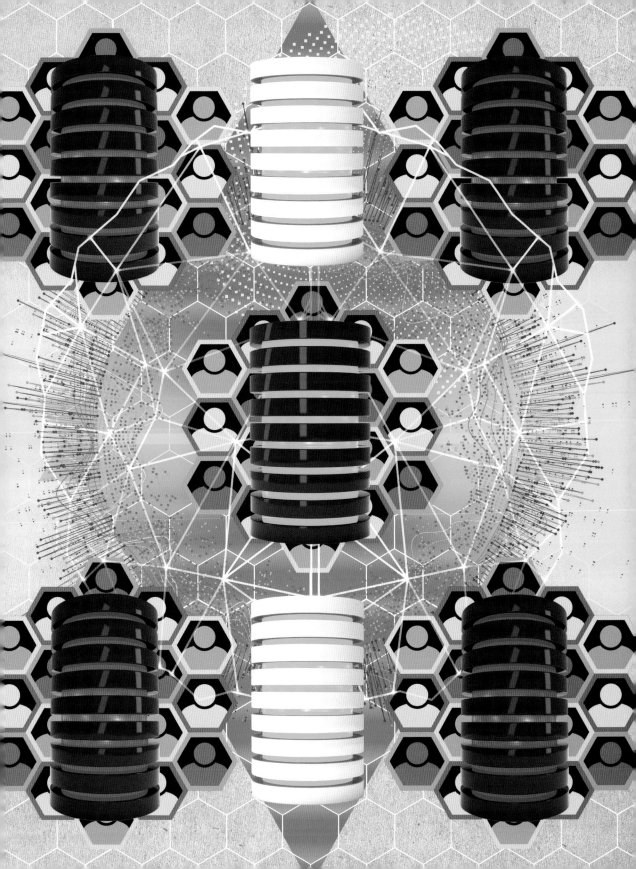

ROBOTS & AI COMPETITIONS

the 30-second data

3-SECOND BYTE
Competitions have fostered scientific and technological progress in AI and robotics since their inception in the 1970s.

3-MINUTE DEEP LEARNING
Successfully deploying robots to tackle a grand challenge leads to significant progress in intelligent robotics, since it typically requires solving several sub-challenges and integrating them to handle problems faced by humans in their daily activities. Examples of these are navigation (the ability for the robot to move around and avoid obstacles), object recognition and tracking (using vision systems to locate and pick up relevant objects), or reasoning about the environment and planning how to act to solve the overall challenge.

Computer board-game competitions have been organized since the 1970s. Computer grand challenges, such as defeating a human board-game world champion, often require some level of machine intelligence. They are used to engage AI researchers in their work; most famous of all is the Turing Test. AI gained media attention after the IBM Deep Blue computer defeated the human chess champion Garry Kasparov in a tournament in 1997. This success inspired AI researchers to launch RoboCup. The challenge is for a team of fully autonomous humanoid robot football players to defeat the human world-champion football team. RoboCup changed the field of AI and robotics by requiring the machine to become a robot – having a body, being able to sense and interact with the physical world and cooperate with other robots. RoboCup paved the way for an explosion of challenges, including the DARPA Grand Challenge, a US contest for autonomous robotic ground vehicles that has advanced the development of autonomous cars and search and rescue humanoid robots. The European Commission funded its first robot competitions to promote research in 2013, leading in 2016 to the European Robotics League (ERL). The ERL challenges focus on civilian problems, and involve robots that collaborate with humans.

RELATED TOPICS
See also
THE TURING TEST
page 34

FAMOUS 'INTELLIGENT' COMPUTERS
page 36

3-SECOND BIOGRAPHIES
DAVID HILBERT
1862–1943
German mathematician who published 23 unsolved problems in mathematics in 1900; they became grand challenges and inspired other similar challenges in science and technology

HIROAKI KITANO
1961–
Japanese scientist, founding president of the RoboCup Federation, co-creator of RoboCup and creator of the AIBO robot dog

30-SECOND TEXT
Pedro U. Lima

Competition is a self-fulfilling prophecy, perhaps creating progress, certainly creating ever more competition.

6 December 1947
Born in Wimbledon, London

1970
Bachelor of Arts in Experimental Psychology (Cambridge University)

1978
PhD in Artificial Intelligence (University of Edinburgh)

1986
Co-authors a seminal paper introducing backpropagation, a central approach in current AI

1987
Joins the University of Toronto (Canada)

1998
Elected Fellow of the Royal Society

2007
Starts collaborating with Google

2012
Co-creates start-up DNNresearch

2013
Joins Google Brain as Google acquires DNNresearch

2017
Co-founds the Canadian Vector Institute for Artificial Intelligence

GEOFFREY HINTON

Geoffrey Hinton is recognized today as the 'godfather of modern AI', but for a long time his research on neural networks was considered 'weak-minded nonsense' by the community of AI scholars. Born in Wimbledon and raised in Bristol, England, his mother was a maths teacher and his father was an entomologist, especially passionate about beetles. His great-great-grandfather was nineteenth-century logician George Boole, the inventor of Boolean algebra, the foundation of modern computing. As a schoolboy, Hinton was not particularly good at maths, but enjoyed physics and football.

He was admitted to the University of Cambridge in 1967 to study physics and chemistry but found the maths in physics too difficult and switched to philosophy and psychology, only to decide that philosophers and psychologists 'didn't have a clue'. He worked as a carpenter for a year before heading to the University of Edinburgh in 1972 to study for a PhD in artificial intelligence under cognitive scientist and chemist Christopher Longuet-Higgins, whose students included Nobel Prize winners John Polanyi and Peter Higgs. Hinton had a 'stormy' student career, since his insistence on researching neural nets was criticized by his supervisor.

In 1987, Hinton established himself in Canada rather than the more popular USA, explaining: 'I didn't want to take money from the US military, and most of the AI funding in the States came from the military.' In 2006, he was experiencing a research crisis due to a lack of funding. But he rebranded his research 'deep learning', a technology based on neural networks – computer programs modelled on the natural neural network of the human brain. He and his team began to beat traditional AI in critical tasks: recognizing speech, characterizing images, and generating natural, readable sentences. In 2009, two of his students won a speech-recognition competition, besting more established methods by using a neural network that was later incorporated into Google's phones.

In 2012, his team won an image-recognition competition using a deep-learning system that involved training a system by using a database of millions of images to recognize and describe an image with a 5 per cent error rate, about the same as humans. The following year, he joined the Google Brain Team, researching how to improve neural network learning techniques. After a life of intellectual perseverance Geoffrey Hinton remains one of the world's star senior researchers in AI.

Luis de Miranda

NANOBOTS

the 30-second data

RELATED TOPICS
See also
GENETIC ENGINEERING
& BIOROBOTICS
page 108

MIND UPLOADING
page 114

MULTI-ROBOT SYSTEMS
& ROBOT SWARMS
page 122

3-SECOND BYTE
Nanobots are robots with sizes in the range of nanometres – such small-scale fabrication is possible through tools such as nanotechnology, DNA manipulation and the use of bacteria.

3-MINUTE DEEP LEARNING
'Gray goo' was a hypothetical apocalypse scenario described by K. Eric Drexler in his 1986 book *Engines of Creation*. Drexler depicted self-replicating nanobots going out of control and consuming all resources to lay waste to the Earth. Although this sounded like science fiction, certain communities were concerned about the possibility of gray goo. In 2004, the British Royal Society published a report on nanoscience stating that such a scenario was implausible.

Until the early twenty-first century, the concept of nanobots remained a hypothetical concept. However, researchers have made technological developments that allow the manufacture of nano-scale machines, just billionths of a metre in size. The most intriguing application area of nanobots is in medicine. Nucleic acid robots, dubbed nubots, are organic molecular 'machines' made up from molecules and proteins within DNA. Through the activation of these DNA structures, nubots can be made to move. Once injected into the body, they may be able to deliver drugs for specific purposes, such as tumour destruction. Another research project has genetically modified Salmonella bacteria to be attracted to tumours. These bacteria can be attached to slightly larger robots carrying drugs, which are released once the tumours are reached. This is another aspect of nanorobotics: the development of micro-scale robots capable of manipulating nano-scale materials. The technology is at a more mature stage than nubots; nano-scale biochips can be manufactured using nanoelectronics, photolithography and biomaterial techniques. Despite the great potential for the use of such technology, there are several risks associated with it. For instance, due to their minuscule size, the micro-scale robots are prone to getting lost within the depths of the human body.

3-SECOND BIOGRAPHIES
RICHARD FEYNMAN
1918–88
American theoretical physicist; the idea of nanobots is attributed to him; in 1959, he described a future with miniaturized machines encoding huge amounts of information

ROBERT FREITAS
1952–
American physicist and nanotechnology scientist known for his contributions to nanomedicine and theoretical discussion about nanorobotics applications

30-SECOND TEXT
Ayse Kucukylimaz

The more we engineer life, the more life will be easily engineered.

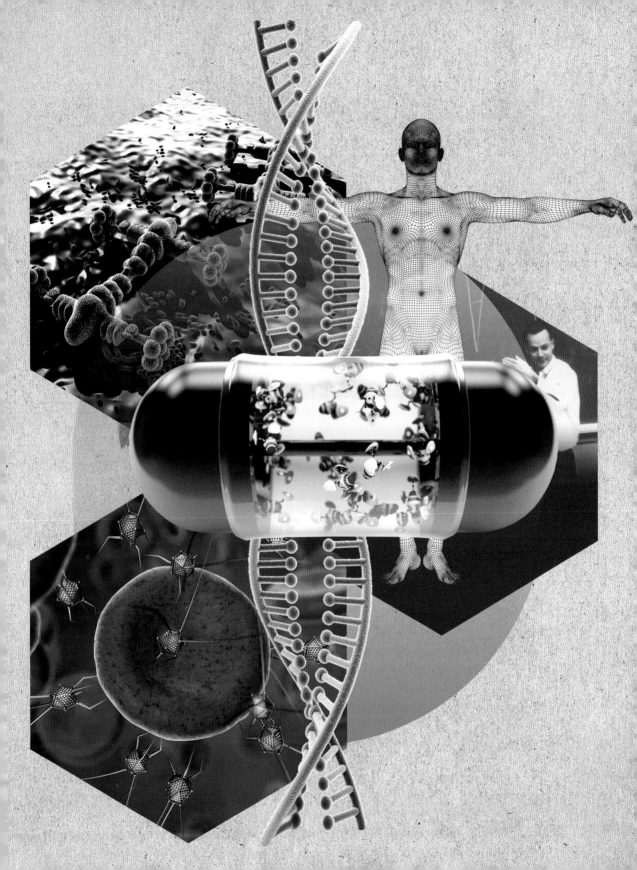

THE INTERNET OF THINGS

the 30-second data

The Internet of Things is a concept
that has pervaded public and private spaces.
An IoT network can be viewed as a distributed
control system, where geographically dispersed
networked sensors acquire information,
communicating with each other via the internet,
to gain awareness of the environment – for
example, the temperature in different areas of
a building, acquired by several sensors. Using
this information, the IoT network can assess
data via the cloud and take action to modify
the state, such as changing the temperature in
different areas of the building. A typical example
is a home where networked sensors capture
temperature and light-intensity values, the
status of household appliances and images of
rooms. Actions can include switching lights on
or off, placing a supermarket order to deliver
goods missing in the fridge, or alerting the
security company because an intruder has been
detected. The settings take into account the
owners' routines and preferences, as well as the
time of day and the weather forecast, acquired
from the internet. Applications cover a wide
range of subjects: smart homes, automated
shops, manufacturing and automation,
environmental monitoring and analysing critical
infrastructure, such as bridges or dams.

RELATED TOPICS

See also
HOME ROBOTS & SMART
HOMES
page 80

COLLECTIVE AI & CLOUD
ROBOTICS
page 124

3-SECOND BYTE

The Internet of Things (IoT)
refers to networks of
physical items connected
to the internet and
equipped with sensors,
actuators, transmitters
and processors for
collecting, processing
and exchanging data.

3-MINUTE DEEP LEARNING

IoT will become increasingly
pervasive in the future
since more powerful
computers will be able to
handle larger amounts of
data; improve predictions,
for example, of potential
traffic congestion; and
enable more sophisticated
actions – for instance,
coordinating the routes of
connected autonomous
cars. On the negative side,
IoT will add to fears about
crime and the invasion of
privacy because it will
become easier to track
people's activities in detail.

3-SECOND BIOGRAPHIES

ROBERT TAYLOR
1932–2017
American technology pioneer
who directed the ARPANET
project. The ARPANET was
a packet-switching network
using the protocol suite TCP/IP
– technologies that laid the
foundations of the internet

KEVIN ASHTON
1968–
British technology pioneer who
named the Internet of Things
in 1999. He later worked to
spread the use of RFID
(radio-frequency identification)
tags, which are used to track
and identify people and objects

30-SECOND TEXT
Pedro U. Lima

*We tend to forget
internet functions in
two directions: we use
it, but it also uses us.*

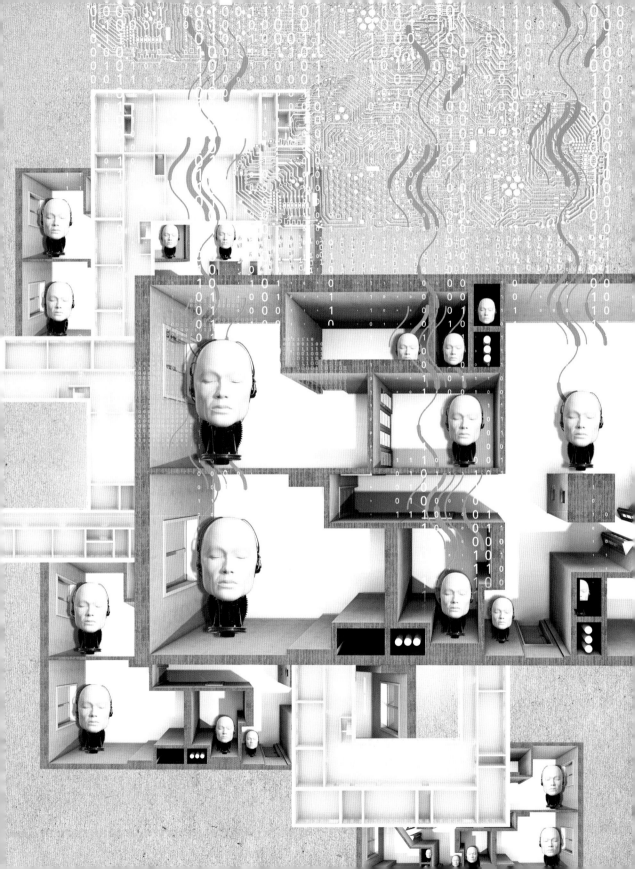

EVOLUTIONARY AI & ROBOTICS

the 30-second data

RELATED TOPICS
See also
MACHINE LEARNING
page 40

RODNEY BROOKS
page 64

THE MERGING OF ROBOTS,
AI & HUMANS?
page 150

3-SECOND BYTE
Darwin's theory about the survival of the fittest is applied to the selection of the most effective algorithms and the best robot-controllers, as if we were gods and machines were becoming alive.

3-MINUTE DEEP LEARNING
The genomes of living organisms are affected by development: events during their lifetime that lead to epigenetic (non-inherited) changes. In biology, this interplay between genetic evolution and epigenetic development is known as 'evo-devo'. Evo-devo emphasizes the importance of environmental factors on an organism's phenotype, the observable aspects of the body. For roboticists, the challenge is to create physically embodied systems that will have enough freedom.

Mimicking nature is a growing trend in robotics and AI. Since most scientists believe nature follows Darwin's principle of evolution by natural selection, they apply this rule to select better algorithms. The theory of evolution states that slight and random changes in an organism's structure can give it an advantage in the wild, under the stress of competition for survival. The advantage allows the individual to survive and to pass on the trait to the next generation. In computer science, this process is called neuroevolution, because machines, unlike humans, are centrally controlled by their artificial central brain. Researchers now use several computers at the same time to create algorithmic competitions. The winning algorithm survives and is adopted while the other poorly performing ones are 'killed'. The principle is the same in evolutionary robotics (ER), where researchers are trying to evolve controllers, which are the microprocessors that control the motors. ER applies the selection, variation and heredity principles of evolution to the design of robots with embodied intelligence. Nobody knows yet if robots will evolve to a point where they will experience perfect synchronization and synergy between their artificial brain and their artificial body. Such an equilibrium has not even been fully achieved in humans.

3-SECOND BIOGRAPHIES
CHARLES DARWIN
1809–82
British naturalist, geologist and botanist, considered to be the founder of the modern science of evolution

LAWRENCE JEROME FOGEL
1928–2007
American pioneer in evolutionary computation. One of the authors of the seminal book *Artificial Intelligence Through Simulated Evolution* in 1966, he undertook pioneering research to generate AI by simulating evolution on computers

30-SECOND TEXT
Luis de Miranda

Computers help us evolve other computers, and thus some programmers play God.

IS THE INTERNET A HIVE MIND?

the 30-second data

3-SECOND BYTE
Thanks to our capacity to be interconnected via the internet, humanity is experiencing a higher level of collective consciousness, for better (wiser insights) and perhaps worse (less diversity).

3-MINUTE DEEP LEARNING
Is the internet like a brain or a mind, or are these just metaphors? Both use electricity and wired networks to fire synapses or send packs of data. But what about consciousness? One hundred years ago, philosopher Teilhard de Chardin coined the term 'noosphere' (from the Greek *knowledge* or *spirit*) to describe the spirit of the Earth as a whole, a 'living tissue of consciousness', growing ever more dense. Some say the internet is to the noosphere what our brain is to our self-awareness.

What is truth? To address this question, most of you won't go for a walk in the forest, but instead google it. The more time we spend on online networks posting our opinions and absorbing ideas and feelings, the more a circle of common influence is created. Overdependence on the internet can create a form of 'groupthink', a standardized way of thinking, and the irrational belief that truth is what the well-connected agree upon. On the positive side, online communication can create a collective awareness about important issues such as a natural disaster. Also, from a sum of organized individuals can come a higher form of understanding, agency or decision making, as in a beehive or a school of fish. Some companies are using human swarms (online groups) to make accurate predictions about the near future, thanks to algorithms that emulate animal swarms but are fed with human input. Just as neurological intelligence is a system of neurons from which an intelligence emerges, smarter than any of the single neurons, human swarm intelligence is a system of brains that work together and are cleverer than any of the individual participants. This is because their collective knowledge, gathered on the internet, can sometimes be synthesized by computers in order to identify a common pattern.

RELATED TOPICS
See also
MULTI-ROBOT SYSTEMS
& ROBOT SWARMS
page 122

COLLECTIVE AI & CLOUD
ROBOTICS
page 124

EXTENDED MINDS
page 146

3-SECOND BIOGRAPHIES
PIERRE TEILHARD DE CHARDIN
1881–1955
French Jesuit priest, palaeontologist and philosopher, he believed that the noosphere, our global consciousness, was evolving towards a higher state

TIM BERNERS-LEE
1955–
British engineer and computer scientist, generally seen as the inventor of the World Wide Web

30-SECOND TEXT
Luis de Miranda

All for internet, and internet for all, so that a choral intelligence might emerge.

WE, ANTHROBOT

WE, ANTHROBOT
GLOSSARY

AI winter Artificial Intelligence is a field that experiences cyclical periods of general interest and funding. In between the periods of enthusiasm for AI are 'wintry' years during which research appears to stagnate. During such periods, it can be beneficial to pause, reflect and generate new ideas.

anthrobotics The idea that we can no longer consider human societies without their interconnections with algorithms and robotic devices. From a collective perspective, we already are a dynamic compound of flesh and protocols, creativity, feelings and electronics. Anthrobotics is a way of looking at the development of AI and robotics from a holistic point of view.

artificial neural networks AI computing systems that are designed to work like a brain. They learn progressively by connecting inputs of information with outputs, through a filtering process that is partly autonomous, and via a collection of nodes called artificial neurons, in which each connection is a simplified version of a synapse. This model is responsible for the success of deep learning since 2011.

big data The act of gathering and storing large amounts of information for analysis or archive. AI uses big data to try to distinguish patterns that humans could not detect.

collective mind and collective consciousness
In 1999, Lawyer Richard Glen Boire proposed the phrase 'virtual collective consciousness' to describe the internet and its capacity to generate a form of collective worldview and behaviour. At the end of the nineteenth century, French sociologist Émile Durkheim spoke of 'collective consciousness' to describe how much our thought is produced by the groups or society we belong to.

Defense Advanced Research Project Agency (DARPA) A national agency connected to the US Department of Defense that invests in long-term research perceived as beneficial for national security. Outcomes have included precision weapons and stealth technology, but also breakthroughs in civilian technologies such as the internet, automated voice recognition and language translation.

symbiosis Literally 'living together': a close and long-lasting relationship between two species. In 1960, computer scientist J. C. R. Licklider published a seminal paper using the metaphor to call for a time when electronic machines and humans would have a symbiotic relationship of benefit to both.

Zeroth Law When two objects are in contact, they exchange heat. The heat flows from the hotter object to the cooler object. At some point, this transfer of heat stops: the objects are in thermal equilibrium. The Zeroth Law of thermodynamics states that any third object in equal equilibrium with one of the first two objects will also be in equilibrium with the second object. In robotics, Zeroth Law is a law introduced by author Isaac Asimov on top of his famous Three Laws of Robotics. It states: 'A robot may not harm humanity, or, by inaction, allow humanity to come to harm.'

INDUSTRY 4.0

the 30-second data

3-SECOND BYTE
Automation enabled the organizational changes that characterized the first three industrial revolutions; AI will do the same for the fourth industrial revolution.

3-MINUTE DEEP LEARNING
Just as the third industrial revolution saw some reduction in physical manufacturing jobs, the fourth may see entirely automated factories. For the first time, the means of production may no longer rely upon human labour. The political implications of this are significant. It seems unlikely that factory owners would choose a human workforce, who require salaries, holidays and pensions, over tireless machines; it is unclear which new jobs may become available instead.

The first industrial revolution is often characterized by the introduction of steam power, used to perform tasks impossible or impractical to perform using human strength alone. While this was indeed revolutionary, the more fundamental change of this industrial revolution (and those that followed) was an organizational one: the bringing together of many workers in one place to help perform a mechanically augmented task. The second industrial revolution was the introduction of the assembly line. This broke the production of complex things into many small tasks performed in sequence as a continuous process. The very first production line, at the Ford Model T factory, reduced the time to produce a car from 12 hours to 90 minutes. As robots began to replace humans in manufacturing, we reached the third industrial revolution – automated factories. The Ford Fiesta factory in Cologne, Germany produces a new car every 86 seconds. The fourth industrial revolution involves logistics. Currently, demand for products is estimated and enough are produced to meet that predicted demand. Excess production means unsold products being scrapped, at financial and environmental cost. Integrating AI systems into the supply chain enables a customer's order to be produced on demand.

RELATED TOPICS
See also
KAREL ČAPEK & THE FIRST 'ROBOTS'
page 18

UNIMATE, THE FIRST INDUSTRIAL ROBOT
page 56

THE MERGING OF ROBOTS, AI & HUMANS?
page 150

3-SECOND BIOGRAPHIES
HENRY FORD
1863–1947
American industrialist who transformed factory production by introducing the assembly line in his Ford car factory

BERTRAND RUSSELL
1872–1970
British philosopher who in 1932 wrote 'In Praise of Idleness', in which he describes the economic impact of automation on a theoretical pin factory

30-SECOND TEXT
David Rickmann

Every new industrial mutation eliminates old jobs and creates new ones – or not.

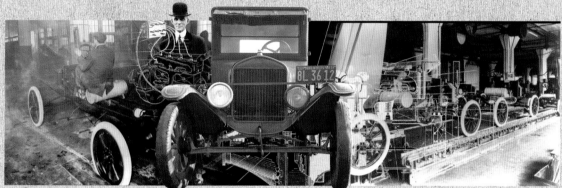

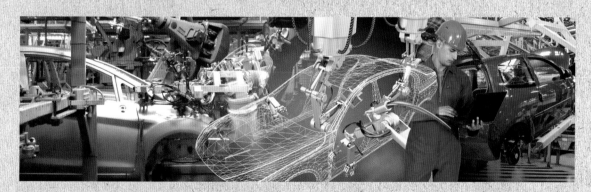

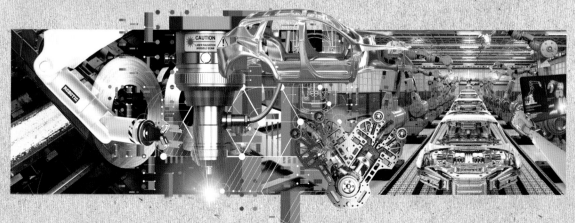

THE THREE LAWS OF ROBOTICS

the 30-second data

3-SECOND BYTE
In 1942 Isaac Asimov proposed three principles that laid the groundwork for robot ethics to this day – but they are not straightforward.

3-MINUTE DEEP LEARNING
Particularly chilling is the Zeroth Law. If averting harm from humanity as a whole justifies every other action, including harming humans, then it would be morally right to kill or enslave individual human beings in order to benefit humanity. We often make choices that are harmful to us or to humanity as a whole – to smoke, to eat sweets, to drive private cars. Should robots be required to prevent us from doing these things?

Isaac Asimov was one of the most prolific science-fiction authors of the twentieth century. Many of his robot stories deal with the problems that arise when robots live and work among humans. Asimov's Three Laws of Robotics were first presented in his short story 'Runaround' (1942). The First Law is 'A robot may not injure a human being, or, through inaction, allow a human being to come to harm.' The Second Law: 'A robot must obey orders given it by human beings, except where such orders would conflict with the First Law.' The Third Law: 'A robot must protect its own existence as long as such protection does not conflict with the First or Second Law.' In 1985, Asimov added the Zeroth Law: 'A robot may not injure humanity, or through inaction, allow humanity to come to harm.' These laws are problematic; it is not clear, for example, what 'harm' means. Giving a student a bad grade is harmful, but is it wrong? Imprisoning criminals harms them, but it is usually seen as necessary and morally right to do so. Sometimes, short-term harm (such as surgery) can prevent long-term damage. Regarding the Second Law, can a robot always be sure which commands would cause long-term benefit or harm?

RELATED TOPICS
See also
ISAAC ASIMOV
page 22

HAL 9000 IN *2001: A SPACE ODYSSEY*
page 24

TERMINATOR & SKYNET
page 28

3-SECOND BIOGRAPHIES
JOHN WOOD CAMPBELL
1910–71
American science-fiction writer/editor who mentored the young Isaac Asimov

ROGER MACBRIDE ALLEN
1957–
American science-fiction author who wrote three novels that use the idea of Asimov's Laws

30-SECOND TEXT
Andreas Matthias

The laws of robotics are meant to be programmed, but their application might lead to judiciary conundrums.

EXTENDED MINDS

the 30-second data

Otto is forgetful. In order to remember anything, he writes it down in a notebook. He always carries his notebook around. When he has to go to an appointment, he looks at the notebook to check where the appointment will take place and how to get there. The notebook thus fulfils some of the functions of Otto's brain. Can we say that Otto's mind is all inside his head, or is the notebook part of it? Clark and Chalmers, in a famous 1998 paper, argue that there is no functional difference between remembering something inside your skull, and remembering it on paper. Whether you add up long numbers in your head or on paper is irrelevant. The mental process of addition is the same, and therefore we can say that mental processes can take place outside a person's head. Today, notebooks are often electronic – the development of AI makes them more accurate than simple reminders. Apps tell us the best way to go to our appointment or which movies to watch, based on our past choices and those of 'people like us'. Big data is like a huge extended memory that shapes our decisions; knowledge is commonly shared and expanded collectively, for example on Wikipedia. What sociologist Emile Durkheim called the 'collective mind' is now a digital reality.

3-SECOND BYTE
Everyday experiences show that our thoughts start partly outside our brain, in our own body, in social bodies or in supports for memory – where do our minds end?

3-MINUTE DEEP LEARNING
The question of responsibility for one's actions is very important, both in criminal law and in case of compensation claims for damages. However, our decisions are increasingly influenced by AI systems that are extremely complex and based on huge amounts of data. It becomes extraordinarily difficult for the end users to question the systems' conclusions; they have to trust them blindly, without being able to exercise their own judgement.

RELATED TOPICS
See also
EMBODIED AI & COGNITION
page 46

HUMAN ENHANCEMENT
page 102

MIND UPLOADING
page 114

3-SECOND BIOGRAPHIES
DAVID EMILE DURKHEIM
1858–1917
French sociologist. One of the fathers of modern social science

HUMBERTO MATURANA
1928–
Chilean biologist who studied how cognition can be defined in terms of the interaction of an organism with its environment

ANDY CLARK
1957–
British philosopher, specializing in the role of the environment in creating mental processes

30-SECOND TEXT
Andreas Matthias

Our multi-connected world is shedding new light on our collective and inter-relational identities.

15 October 1929
Born in Indiana, USA

1947
Begins studying
philosophy at Harvard

1952–64
Extended study trips
and research fellowships
in Europe

1957–59
Instructor in Philosophy
at Brandeis University,
Massachusetts

1960–68
Teaches at the
Massachusetts Institute
of Technology

1968
Joins UC Berkeley as
an associate professor
of philosophy

1972
Publishes *What
Computers Can't Do:
A Critique of Artificial
Reason*

1986
Publishes *Mind Over
Machine: The Power of
Human Intuition and
Expertise in the Era of
the Computer*, with his
brother Stuart Dreyfus

2001
Fellow at the American
Academy of Arts and
Sciences

2010
Documentary movie
Being in the World, based
on Dreyfus' interpretation
of Heidegger

22 April 2017
Dies at the age of 87

HUBERT DREYFUS

Born in Terre Haute, Indiana,
USA, Dreyfus was educated at Harvard
University. During the unusually long years
between achieving his master's degree in
1952 and his PhD in 1964, Dreyfus studied
continental philosophers in Europe. He met
with some of the greatest philosophers of the
twentieth century: Martin Heidegger, Jean-Paul
Sartre and Maurice Merleau-Ponty. But he was
left unimpressed, and preferred to study these
philosophers through their writings.

Dreyfus' early years as a philosopher were,
at the same time, the early years of AI research.
From Alan Turing's paper in 1950 to Newell and
Simon's Physical Symbol System Hypothesis
(1976), early AI researchers saw intelligence as
a process of abstract symbol manipulation,
like a game of chess or a formal deduction in
a logic calculus. But Dreyfus always opposed
this view. Based on arguments from Heidegger
and the so-called 'continental' tradition in
philosophy, he believed that what makes us
human is not abstract, formal intelligence,
but the freedom we have to define ourselves
through the ways we relate to the world. For
Dreyfus, 'skilful coping' with the environment
became the hallmark of human intelligence.

In his influential criticism of expert systems,
he argued that human expertise, unlike that
of software programs, is not merely an ability
to follow rules. A human expert, such as an
expert chess player, is not simply someone
who can calculate their moves more quickly
than their opponent (as a computer would do).
Instead, developing expertise is a long process,
in which the human brain learns to recognize
particular features of the board, and to relate
these features to possible reply moves. The
more skilled the expert, the more inexplicable
their moves will be, taking into account very
subtle features of the overall position on the
board. Expert play is very far removed from
the purely 'calculative' approach of the
beginner player. The same applies, Dreyfus
thought, to any acquired skill, such as driving
a car, teaching or cooking.

In this way, Dreyfus developed a deep respect
for all human skills. In his lectures and books,
he regularly applied his theories to all areas of
everyday life, from engineering to religious
questions and to the search for meaning in
human life. History has shown him to be
correct. Symbolic AI never became really
successful. After the 'AI winter' of the 1980s,
it was superseded by neural networks and
deep learning, methods of machine learning
that are more biologically inspired and not
based on abstract sets of rules.

Andreas Matthias

THE MERGING OF ROBOTS, AI & HUMANS?

the 30-second data

RELATED TOPICS
See also
HAL 9000 IN *2001: A SPACE ODYSSEY*
page 24

HUMANOID ROBOTS
page 90

TRANSHUMANISM & SINGULARITY
page 116

3-SECOND BYTE
Once a science-fiction hypothesis, the idea that humans and intelligent machines are blending into a new species is taken seriously even by official institutions – but is this a novel theory?

3-MINUTE DEEP LEARNING
In *The Myth of The Machine*, Lewis Mumford showed how civilized societies functioned somewhat like gigantic machines. Humans are often controlled by protocols, and institutions use them to perform procedures, as if they were machine cogs. Technology enables us to do things we could not do otherwise. But enablement, in psychology studies, is a term that also designates dependency and the risk of addiction. Are we using machines or are social machines using us?

Even the official US Defense Advanced Research Project Agency (DARPA) has declared that it believes the merging of humans and machines is happening now, crossing a new technological boundary. The power of computers and AI allows us to test our ideas better and faster. Robotic prostheses are making such progress that people are able to regain usage of their members, control machines via brain implants or walk anew thanks to exoskeletons. The symbiosis of computers and humans announced by computer scientist J. C. R. Licklider in 1960 seems to be happening in front of our eyes. Consider how lost and powerless we feel when we don't have an internet connection! Technology is not only about tools that help us, but it is also a part of who we are. We are becoming anthrobots, an intertwined combination of flesh and artificial systems. But is this really a new paradigm? In fact, the human species can be defined by its continual adoption of new systems since the beginning of history. Language and grammar, for example, allowed humans to communicate so that they could extend their will. The division of labour enabled us to create human machines of workers performing mechanical tasks. Robots are fascinating because we recognize ourselves in them.

3-SECOND BIOGRAPHIES
LEWIS MUMFORD
1895–1990
American historian and philosopher who criticized the tendency of 'megatechnical' societies to create deep human dissatisfaction, because selfhood seems irrelevant to efficient systems

PIERRE BOURDIEU
1930–2002
French sociologist who showed how even our most intimate tastes and preferences are largely determined by the social group we belong to

30-SECOND TEXT
Luis de Miranda

We are already an anthrobotic species, living in symbiosis with machines.

APPENDICES

RESOURCES

NON-FICTION BOOKS

Advances in Culturally-Aware Intelligent Systems and in Cross-Cultural Psychological Studies
Edited by Collette Faucher
(Springer, 2018)

Anatomy of a Robot: Literature, Cinema, and the Cultural Work of Artificial People
Despina Kakoudaki
(Rutgers University Press, 2014)

Army of None: Autonomous Weapons and the Future of War
Paul Scharre
(W. W. Norton & Company, 2018)

Artificial Intelligence: A Modern Approach
Stuart Russell and Peter Norvig
(Pearson, 2016)

Deep Learning
Ian Goodfellow, Yoshua Bengio, Aaron Courville
(MIT Press, 2017)

Driverless Cars: On a Road to Nowhere
Christian Wolmar
(London Publishing Partnership, 2018)

The Fourth Education Revolution: How Artificial Intelligence is Changing the Face of Learning
Anthony Seldon
(The University of Buckingham Press, 2018)

The Glass Cage: Automation and Us
N. Carr
(W. W. Norton & Company, 2014)

The Internet of Things
Samuel Greengard
(MIT Press, 2015)

The Living Brain
G. Walter
(W. W. Norton Company, Inc., 1953)

The Master Algorithm
Pedro Domingos
(Penguin Books, 2017)

Robots in Space: Technology, Evolution, and Interplanetary Travel
Roger D. Launius and Howard E. McCurdy
(The Johns Hopkins University Press, 2012)

Social Robots: Boundaries, Potential, Challenges
Marco Nørskov
(Routledge, 2015)

FICTION BOOKS

Accelerando
Charles Stross
(Orbit, 2006)

I, Robot
Isaac Asimov
Gnome Press (1950)

Neuromancer
William Gibson
(Ace Books, 1984)

MAGAZINES/JOURNALS

AI & Society
link.springer.com/journal/146

Ethics and Information Technology
www.springer.com/computer/swe/journal/10676

Futures
www.journals.elsevier.com/futures

WEB SITES

Bristol Robotics Laboratory
www.brl.ac.uk/researchthemes/swarmrobotics.
aspx
Swarm robotics projects – a new approach
to coordinate the behaviours of a large number
of relatively simple robots in a decentralized
manner.

DeepMind
deepmind.com/research/alphago
Company dedicated to artificial intelligence
research.

European Robotics League
eu-robotics.net/robotics_league
The home of two indoor robotics competitions.

Future Technologies and Their Effect
on Society – Nanobots
futureforall.org/nanotechnology/nanobots.html
Explanation of nanobots and links to articles.

Futurism
futurism.com
Includes news about AI, earth and energy,
health and medicine and space.

Machina Speculatrix
www.extremenxt.com/walter.htm
About Grey Walter's turtle-like mobile
robotic vehicles.

RoboCup Federation official website
www.robocup.org
Website of the World Championship
on robotics.

SingularityHub
singularityhub.com
Focused on AI in robotics, neuroscience,
computing, biotech and more.

War on the Rocks
warontherocks.com
Includes news about military AI.

What is Deep Learning?
machinelearningmastery.com/what-is-deep-
learning
A range of experts and leaders in the field
explain deep learning.

NOTES ON CONTRIBUTORS

EDITOR

Luis de Miranda, PhD (University of Edinburgh), is a philosopher and historian of ideas who has carried out research into anthrobotics, digital cultures and how technology is enmeshed with our everyday life. He is the author of several fiction as well as non-fiction books, for example *L'Art d'être libres au temps des automates* ('The Art of Freedom in the Time of Automata'). He works on AI and digital humanities at Örebro University, Sweden.

CONTRIBUTORS

Sofia Ceppi is an academic researcher in Artificial Intelligence with a Computer Science and Information Technology Engineering background. She spent most of her career studying how machines and humans can interact, collaborate and influence or incentivize each other.

Neha Khetrapal is a researcher who investigated language development in autism for her PhD at Macquarie University. Apart from academic research and teaching, she is also involved with developing applications for rehabilitation of children with autism and other outreach activities for the benefit of children with special needs.

Ayse Kucukyilmaz is a lecturer in Computer Science, and a member of the Lincoln Centre for Autonomous Systems (L-CAS). Her research interests include haptics, human-robot interaction and machine learning applications in adaptive autonomous systems.

Pedro U. Lima is a Professor at IST, University of Lisbon, where he teaches and carries out research on Intelligent Robots and Autonomous Systems. He is a Trustee of the RoboCup Federation and has organized several robotics dissemination events in Europe and Portugal, including the Portuguese Robotics Open since 2001 and RoboCup 2004.

Andreas Matthias teaches philosophy at Lingnan University, Hong Kong. He specializes in AI and robot ethics. For twenty years before that, he was a professional software developer. His publications include the articles 'Robot Lies in Health Care. When Is Deception Morally Permissible?' (2015), 'Algorithmic moral control of war robots: Philosophical questions' (2015) and the book *Automata as Holders of Rights. A Proposal for a Change in Legislation* (2008).

Lisa McNulty is an academic philosopher, and a Fellow of the Philosophy of Education Research Centre at Roehampton University. She has research interests in social epistemology, and has extensive teaching experience in the area of applied ethics.

David Rickmann is a roboticist and transport consultant specializing in modelling, simulations and the application of large-scale data analysis to transport issues. He has worked in all areas of transport robotics including car factory robots, autonomous trains and sensor systems for highways.

Mario Verdicchio is an academic researcher from Milan, Italy. He obtained a PhD in Information Engineering at Politecnico di Milano, Italy in 2004 with a thesis on software agents. Since then, he has been pursuing several research and educational projects in academia with a focus on how computer technology and human endeavours interact. He co-founded the xCoAx international conference series on computer art.

INDEX

ACKNOWLEDGEMENTS

The publisher would like to thank the following for permission to reproduce copyright material on the following pages:

Alamy/Everett Collection, Inc: 127TL; Pictorial Press Ltd: 19C; Science History Images: 38; Chris Wilson: 88.
Andrew Averkin (www.andrewaverkin.com): Front cover.
Courtesy Berkley, University of California © 2018 UC Regents, all rights reserved: 148.
Bibliothèque nationale de France: 17CR (BG).
DARPA: 67C.
Flickr/OnInnovation: 110; TechCrunch: 65.
Getty Images/Bettmann: 95C, 103C; Daily Herald Archive: 97C; General Photographic Agency: 127TR; Hulton Archive/Stringer: 2; Brendan Hunter: 105C; Keystone Features: 9; Keystone-France: 147C; John Pratt/Stringer: 27CR; Richvintage: 113C; Universal Images Group: 85C.
Library of Congress, Washington DC: 15CL, 15CR, 19BG, 22, 143C.
LSST Project/J.Andrew: 71C.
NASA: 109C; JPL-Caltech: 71C; JPL-Caltech/Cornell: 71C (BG).
Shutterstock/3000ad: 7C (BG), 25C (BG), 151TL, 151CL, 151TL; 3d_kot: 81C; 3Dsculptor: 61C (BG); Adike: 117CL; Agsandrew: 29BG, 45C; Aha-Soft: 122C; Alejo Miranda: 45TC; Aleksandrs Bondars: 137C; Alex Staroseltsev: 145C; Alexandra Shargaeva: 35C; Alexdndz: 87C (BG); AlexLMX: 87C; Alexwhite: 97C; Aleynikov Pavel: 21BG; all_about_people: 95C (BG); all_is_magic: 97BG, 135C; and4me: 087TR & TL (BG); andrea crisante: 21C (BG); Andrey Suslov: 87C (BG), 115TC; andrey_l: 151TR; Antiv: 107CR (BG), 115TC; Anton Zabieiskyi: 117CL; Aphelleon: 7C, 25C, 29BC; April Cat: 37C; Archideaphoto: 127C; archy13: 29BC; art-sonik: 43C; artellia: 107BC; artemiya: 59C; ArtHead: 41C (BG); Ase: 107C; Asharkyu: 57C; Axstokes: 59BC; Baivector: 35C (BG); BallBall14: 73CL & CR; Baloncici: 85BC; Baoyan: 91CL; BH007: 37TC, 37C; Blackdogvfx: 103C; Blank-k: 87CL & CR; Bloody Alice: 29BG; Brovkin: 125BG; Brovko Serhii: 127BC; Carolina K. Smith MD: 59BG; Chaikom: 27C (BG); Cherezoff: 133; Chesky: 81C, 143BCL; Clash_Genre: 91TC; Cliparea|Custom Media: 35CR & CL; COLOA Studio: 49CL; Daboost: 147TR; DarkGeometryStudios: 29C; Darsi: 45TC; David Petrik: 73TC; design36: 93CR, 107CL, 131TR; DG-Studio: 93BG; Di Ma: 69C (BG); Diuno: 29C (BG); DivinHX: 91C; Djem: 109C (BG); DM7: 151C; Dmitriy Rybin: 21BC; donatas1205: 17TR, 17BC, 17BG; dreamer: 107C (BG); DRN Studio: 127BR; Elnur: 63C; Evannovostro: 61C; Everett Collection: 87C, 95C; Everett Historical: 69TC; Executioner: 43C; Exkluzive: 47TL & CR; Express Vectors: 73C (BG), 85TR (BG); F.Schmidt: 59TC; Filip Warulik: 35C; Flowgraph: 107CR; Forest Foxy: 87BG; Franke de Jong: 122TR; G_O_S: 73CL & CR; ga-ko: 67C (BG); GarryKillian: 7BG, 25BG, 61BG, 122BG, 125BG; Ggaryfox45114: 107C; GiroScience: 109C; GlebSStock: 115BC; Goran Bogicevic: 73C; Gouraud Studio: 113BG; graphicINmotion: 91C; Gumenyuk Dmitriy: 49C, 49CR; Gun2becontinued: 27C (BG); Gzibon: 143CL; Haver: 147C; Hkeita: 113CL; HQuality: 41C; Iaremenko Sergii: 109C; Iaroslav Neliubov: 57CL & CR, 61C, 115TC, 143BC; icon99: 81C; iconvectorstock: 73C (BG); Igor Kisselev: 125BG; Igor Marusichenko: 49BR; Imredesiuk: 131TL; Ink Drop: 49BG; Iosif Tchibalashvili: 21C (BG); Iruii: 7C, 25C, 107C (BG); lunewind: 127C; Jackie Niam: 51C (BG), 85C (BG), 117BG; James Steidl: 19C; Jan Schneckenhaus: 87C; Jaromir Chalabala: 143BCL; Jelome: 43C; jm1366: 51C (BG); Jolygon: 63CR; Jozsef Bagota: 87C (BG); Jut: 137C; KindheartedStock: 91C (BG), 125BG; koya979: 109C; Krahovnet: 63C

(BG), 151BG; Kristina Shevchenko: 113BR; Ktsdesign: 137C; Kurhan: 115C; Kvsan: 57C; LazyLuck: 45C; IgorZh: 29BG, 29C (BG); Linda Bucklin: 135C; Liu; zishan: 47C (BG); Liya Graphics: 131C; Ljupco Smokovski: 85C, Login: 82C, Markus Gann: 97BG, 147C; Marysuperstudio: 43CL; meow_meow: 51BG; Mertsaloff: 17BG; Milart: 43BG; Millissa4like: 82C (BG); Miloje: 27CR (BG); Mitoria: 43BC; Mix3r: 82BG; Monopoly919: 59BG; Mopic: 105C (BG); Morphart Creation: 17C (BG); NASA Images: 71BG, 113TC, 151T; Natykach Nataliia: 103BG; Newelle: 49BG; Nikelser: 49BG; Ninell: 7BC (BG), 25BC (BG); Nmid: 131C; Nobeastsofierce: 131BL; Norrapat Thepnarin: 27C (BG); Nostalgia for; Infinity: 71TR; Ociacia; 47BC, 61C, 117CR, 151C, 151C; Octomesecam: 45C (BG); Oleksiy Mark: 93C (BG), 115C (BG), 143BC; Omelchenko: 59BG, 078BG; Ozz Design: 41CR; Pakawat Suwannaket: 95BG; Paper Street Design: 63C (BG), 93CL; Patryk Kosmider: 85C (BG); Petr Bonek: 15BC; Phonlamai Photo: 35CR, 47TC, 135CR, 143BCR; photoart985: 7C, 25C; PhotoStock10: 143BL; Phovoir: 143BC; Piotr Adamowicz: 73C; pixeldreams.eu: 43TC; pixelparticle: 29BG; pixidsgn: 115C; posteriori: 122C; PP77LSK: 57C, 143BC; Prapaporn Sonrach: 113BG, 113BL; pro500: 117C; Pumal Vittayanukorn: 115TC; Quaoar: 15TR; R Mendoza: 51C (BG); R.T. Wholstadter: 7C (BG), 25C (BG); Rangizzz: 37TL; Romanova Natali: 115C (BG); Romolo Tavani: 7BC, 25BC; Ryzhi: 61C; Sabphoto: 29BG; Sakkmesterke: 21BG; Sarah Holmlund: 051C, 133, 145BC; Sattahipbeach: 47BG; Sdecoret: 29T (BG); Sebastian Kaulitzki: 63CL; Sergey Tarasov: 147C; SergyBitos: 143C; ShadeDesign: 41C; ShaunV: 45TC; SkillUp: 41C (BG); Slavoljub Pantelic: 69C, 69T; Somachai Som: 73BC (BG); Sonja Filitz: 122TR; Sportpoint: 105CR; Stationidea: 45C; Steroids: 37C (BG); Studi8Neosiam: 59C; Studiovin: 57C; Subbotina Anna: 151C; Sumkinn: 27C (BG); Supergenjijalac: 143BCR; Sylverarts Vecros: 133TR; Tatiana Shepeleva: 35CL, 47C; TeddyGraphics: 113CR; Telesniuk: 145TR; Tonis Pan: 147BC; Triff: 95C (BG); Umberto Shtanzman: 147C; Vadim Sadovski: 7BG, 25BG; Vector Plus Image: 45C; Vector.design: 95CL & CR; Victor Habbick: 45CL & CR; Vikpit: 105C (BG); Viktorus: 35BG; Visaro: 82C; Visual Generation: 73C & CR; Vitstudio: 103BG, 107C (BG); Vladystock: 47C, 51C (BG), 93BG; Vshivkova: 109C (BG); Will Thomass: 37C; Willyam Bradberry: 78C, 151C; XiXinXing: 145TC; Your: 059C (BG); Zack Frank: 67C; Zapp2Photo: 63BC; Zerbor: 37BR; Zoltan Pataki: 45TC.
Courtesy University of Toronto/Johnny Guatto: 128.
US Navy/Mass Communication Specialist 1st Class James E. Foehl: 67CL (BG).
United States Department of Energy: 131CR.
Wikimedia Commons/Anagoria: 143TC; Cyberdyne Studio/Yuichiro C. Katsumoto: 105BL; Mario Del Curto: 17C; Fashadarvin: 122C; GillyBerlin: 143BC; Manuel Kehrli: 17CR; Kriplozoik: 15TR; David Monniaux: 37CR; Alison Pope: 21C; Anders Sandberg: 27CL.
Zuiden, Fotopersbureau Het: 78CR.

All reasonable efforts have been made to trace copyright holders and to obtain their permission for the use of copyright material. The publisher apologizes for any errors or omissions in the list above and will gratefully incorporate any corrections in future reprints if notified.